ADVANCES IN CHEMICAL PHYSICS

VOLUME 144

ADVANCES IN CHEMICAL PHYSICS

VOLUME 144

Series Editor

STUART A. RICE

Department of Chemistry
and
The James Franck Institute
The University of Chicago
Chicago, Illinois

A JOHN WILEY & SONS, INC., PUBLICATION

Published by John Wiley & Sons, Inc., Hoboken, New Jersey
Published simultaneously in Canada

For general information on our other products and services or for technical support, please contact our Customer Care Department within the United States at (800) 762-2974, outside the United States at (317) 572-3993 or fax (317) 572-4002.

Wiley also publishes its books in a variety of electronic formats. Some content that appears in print may not be available in electronic formats. For more information about Wiley products, visit our web site at www.wiley.com.

Library of Congress Catalog Number: 58-9935

ISBN: 978-0470-54786-1

Printed in the United States of America

10 9 8 7 6 5 4 3 2 1

CONTRIBUTORS TO VOLUME 144

JAMES F. LUTSKO, Center for Nonlinear Phenomena and Complex Systems CP 231, Universite Libre de Bruxelles, 1050 Brussels, Belgium

KAZUO TAKATSUKA, Department of Basic Science, Graduate School of Arts and Sciences, University of Tokyo, Komaba, 153-8902, Tokyo, Japan

TAKEHIRO YONEHARA, Department of Basic Science, Graduate School of Arts and Sciences, University of Tokyo, Komaba, 153-8902, Tokyo, Japan

J. STECKI, Department III, Institute of Physical Chemistry, Polish Academy of Sciences, Warsaw, Poland

INTRODUCTION

Few of us can any longer keep up with the flood pf scientific literature, even in specialized subfields. Any attempt to do more and be broadly educated with respect to a large domain of science has the appearance of tilting at windmills. Yet the synthesis of ideas drawn from different subjects into new, powerful, general concepts is as valuable as ever, and the desire to remain educated persists in all scientists. This series, *Advances in Chemical Physics*, is devoted to helping the reader obtain general information about a wide variety of topics in chemical physics, a field that we interpret very broadly. Our intent is to have experts present comprehensive analyses of subjects of interest and to encourage the expression of individual points of view. We hope that this approach to the presentation of an overview of a subject will both stimulate new research and serve as a personalized learning text for beginners in a field.

STUART A. RICE

CONTENTS

RECENT DEVELOPMENTS IN CLASSICAL DENSITY FUNCTIONAL THEORY

JAMES F. LUTSKO

Center for Nonlinear Phenomena and Complex Systems CP 231, Université Libre de Bruxelles, 1050 Brussels, Belgium

CONTENTS

Advances in Chemical Physics, Volume 144, edited by Stuart A. Rice
Copyright © 2010 John Wiley & Sons, Inc.

I. INTRODUCTION

The central question in equilibrium statistical mechanics is the calculation of various physical quantities—pressure, magnetization, charge distribution, and so on—from the known many-body distribution function. The technical difficulty lies in first formulating the macroscopic quantity as the average of some quantity and then performing the calculation. The first task is often straightforward. In most cases, the second task can be cast as an average over the one- or two-body distributions that result from integrating the original N-body distribution over N-1 or N-2 of the coordinates. In equilibrium, the velocity distribution is always Maxwellian and therefore trivial, so the real work concerns the configurational part of the averages. The configurational part of the one-body distribution is precisely the same as the average local (number) density, while that of the two-body distribution is closely related to the pair distribution function. As explained below, the pair-distribution function can itself be viewed as the one-body distribution in a system subject to a particular external potential, so that it follows that a large part of equilibrium statistical mechanics is solved by a general method to obtain the one-body distribution—or, equivalently, the local density—of systems subject to arbitrary external fields. This is the rationale behind Density Functional Theory (DFT), which aims to provide just such a method.

    The modern approach to DFT can be traced back to the work of van der Waals on the free energy of fluids made inhomogeneous by gravity [1, 2]. Consider a system confined to a volume V in which the local density is $\rho(\mathbf{r})$ and the average density is $\bar{\rho} = \int_V \rho(\mathbf{r})\, d\mathbf{r}$. Naively, one might imagine that if the free energy per unit volume in a bulk fluid, for which $\rho(\mathbf{r}) = \bar{\rho}$, is $f(\bar{\rho})$, so that the total free energy $F = Vf(\bar{\rho})$, then the free energy of the inhomogeneous fluid is $F = \int_V f(\rho(\mathbf{r}))\, d\mathbf{r}$. However, as discussed by van der Waals, a small volume of fluid bounded on one side by fluid at a lower density and on the other side by fluid at a higher density will feel a force due to the different numbers of interactions between it and the neighboring volumes. Thus, the free energy must contain terms taking this into account and a simple analysis of the forces leads to squared-gradient terms, thus giving the well-known squared-gradient model, which is still widely used today. Similar ideas were rediscovered in the form of Ginzburg–Landau theory [3] and in the work of Cahn and Hilliard [4] on planar interfaces.

However, DFT is more than just the idea that the free energy can be expressed as a functional of a set of order parameters. The formal development of DFT begins with the theorem of Mermin that, for a given temperature and chemical potential, an external field will give rise to a unique equilibrium density distribution and that this profile will minimize a particular functional [5]. The uniqueness of the external field/ density distribution mapping means that the field can be eliminated in favor of the density so that the functional determining the density has only a trivial dependence on the field. Once the field-independent, nontrivial part of that functional is known, *all* problems involving external fields can in principle be solved.

The required functional can furthermore be expressed, exactly, as an infinite series constructed from the correlation functions of a bulk homogeneous fluid. Since liquid-state theory gives models for those functions, the earliest approaches to DFT involved trying to use the lowest terms from that series as an approximate model. This led to the first DFT of freezing by Ramakrishnan and Yussouff [6] as well as to the work by Saam and Ebner on the properties of nonuniform fluids such as the liquid–vapor interface [7, 8]. Many other theories followed, all based on the idea of using knowledge of bulk liquids to model the DFT functional for more complex systems.

At about the same time that this early work was beginning, in the mid-1970s, Percus gave the exact DFT for one-dimensional hard rods [9]. Of course, attempts were quickly made to generalize these results to higher dimension with limited success. However, in the 1990s, Rosenfeld formulated Fundamental Measure Theory (FMT) as a generalization of the old Scaled Particle Theory [10–12] and it was soon recognized that this was best thought of in terms of a generalization of Percus' results. FMT represented a departure from other methods in that it does not explicitly depend on some sort of mapping to an effective liquid. As discussed below, it has many attractive features and is currently considered the best approximate DFT for hard-sphere systems.

Implicit in all of this is the fact that hard-core systems are much better understood than are more realistic particles with attractive interactions. However, as is the case in liquid-state theory, the hard-core system is a useful first approximation and attractive interactions can be treated in a perturbative or mean-field fashion. This represents the state of the art in most calculations. On this basis, the use of DFT has exploded so that today, new papers based on DFT are published in numerous journals every month.

Over the course of the development of the history, many important reviews have been written. The 1979 review by Evans [13] helped to unify and define the subject, and his 1992 review [14] is an excellent summary of the state of DFT at that time, including a very interesting discussion of the history of DFT and the interplay between the classical and quantum theories. Important overviews have also been written by Baus and Lutsko [15] and Löwen [16]. An excellent review stressing applications is given by Wu [17]. The point of the present chapter is to describe DFT in its current form. The next section gives the formal theory. It seems appropriate to emphasize this because it is important to understand the limitations of the theory as well as its use. Section III presents a review of liquid-based DFT models. While FMT is viewed as the most accurate model for hard-core systems, it is also more complicated than the earlier theories and computationally more demanding. Thus, it is still quite common to see even the simplest DFT models used (e.g., in the study of three-body contributions to freezing [18] and of glasses [19]) while, for example, the Modified Weighted Density Approximation (discussed below) is also still frequently used [20]. The next section is devoted to explaining the development of FMT starting with the exact results of Percus, the SPT-inspired FMT of Rosenfeld followed by the important idea of dimensional reduction and culminating in the very accurate functionals currently in use. Notwithstanding its successes, FMT is not perfect and the section concludes with a discussion of open questions and problematic issues. Section V addresses the problem of attractive interactions and gives some representative calculations for simple fluids. Section VI is devoted to attempts to use DFT to understand dynamical phenomena, the so-called Dynamical Density Functional Theory and Energy Surface methods. The chapter ends with a short summary.

II. FUNDAMENTALS

A. Statistical Mechanical Preliminaries

Since the local density is the fundamental unknown, DFT is formulated in the grand canonical ensemble with the temperature T and the chemical potential μ as fixed parameters. Consider a classical, conservative system of N particles having coordinates and velocities, $\mathbf{q}_i$ and $\mathbf{p}_i$ respectively. The dynamics are governed by a Hamiltonian $\hat{H}_N = \hat{K}_N + \hat{V}_N + \hat{U}_N,$ where the kinetic energy is

$$\hat{K}_N = \sum_{i=1}^{N} p_i^2 / 2m_i \tag{1}$$

The potential is an arbitrary function of the coordinates, $\hat{V}_N = V(\mathbf{q}_1, \ldots, \mathbf{q}_N)$ and the energy of interaction with the external field is

$$\hat{U}_N = \sum_{i=1}^{N} \phi(\mathbf{q}_i) \tag{2}$$

In the following, a caret will indicate a microscopic quantity having an instantaneous value calculated from the particle positions and momenta. The equilibrium distribution, f, giving the probability density that the system consists of N particles with positions $\mathbf{q}^N \equiv \mathbf{q}_1, \ldots, \mathbf{q}_N$ and momenta $\mathbf{p}^N \equiv \mathbf{p}_1, \ldots, \mathbf{p}_N$ is

$$f(\mathbf{q}^N, \mathbf{p}^N, N; [\phi]) = \frac{1}{\Xi[\phi]N!h^{DN}} \exp(-\beta(\hat{H}_N - \mu N)) \tag{3}$$

where the inverse temperature is $\beta = 1/k_B T$, k_B is Boltzmann's constant and h is Planck's constant. The notation $[\phi]$ on the left-hand side indicates a *functional* dependence. The grand potential, Ω, and the grand partition function, Ξ, are

$$\Xi[\phi] = \exp(-\beta\Omega[\phi]) = \sum_{N=0}^{\infty} \frac{1}{N!h^{DN}} \int \exp(-\beta(\hat{H}_N - \mu N)) d\mathbf{p}^N d\mathbf{q}^N \tag{4}$$

where $d\mathbf{q}^N = d\mathbf{q}_1 \ldots d\mathbf{q}_N$, and so on.

The local number density simply counts the number of particles in a given volume and so its microscopic expression is

$$\hat{\rho}(\mathbf{r}) = \sum_{i=1}^{N} \delta(r - \mathbf{q}_i) \tag{5}$$

Note that we use the traditional notation even though in other contexts the symbol ρ more commonly refers to the mass density. For single-component systems, this is unimportant but care must be exercised when the particles have different masses. The contribution of the external field to the Hamiltonian can be written as

$$\hat{U}_N = \int \phi(\mathbf{r})\hat{\rho}(\mathbf{r})d\mathbf{r} \tag{6}$$

Hence, it immediately follows that

$$\frac{\delta\Omega[\phi]}{\delta\phi(\mathbf{r})} = \Xi^{-1} \sum_{N=0}^{\infty} \frac{1}{N!h^{DN}} \exp(\beta\mu N) \int \hat{\rho}^N(\mathbf{r}) \exp(-\beta\hat{H}_N) d\mathbf{p}^N d\mathbf{q}^N$$

$$= -\langle \hat{\rho}(\mathbf{r}) \rangle$$

$$\equiv -\rho(\mathbf{r}; [\phi]) \tag{7}$$

where $\langle \ldots \rangle_\phi$ denotes an average in the grand canonical ensemble with field ϕ and the last line defines the average local equilibrium density, $\rho(\mathbf{r}; [\phi])$. From its definition, it is easy to see that $\rho(\mathbf{r}; [\phi])$ is the probability to find a particle at position $\mathbf{r}$. A second functional derivative gives

$$\frac{\delta^2 \Omega[\phi]}{\delta\phi(\mathbf{r}_1)\delta\beta\phi(\mathbf{r}_2)} = -\frac{\delta\rho(\mathbf{r}_1; [\phi])}{\delta\beta\phi(\mathbf{r}_2)}$$

$$= \langle \hat{\rho}(\mathbf{r}_1)\hat{\rho}(\mathbf{r}_2) \rangle - \langle \hat{\rho}(\mathbf{r}_1) \rangle \langle \hat{\rho}(\mathbf{r}_2) \rangle$$

$$= \rho(\mathbf{r}_1)\delta(\mathbf{r}_1 - \mathbf{r}_2) + \rho(\mathbf{r}_1)\rho(\mathbf{r}_2)h(\mathbf{r}_1, \mathbf{r}_2; [\phi]) \qquad (8)$$

where the structure function $h(\mathbf{r}_1, \mathbf{r}_2; [\phi]) = g(\mathbf{r}_1, \mathbf{r}_2; [\phi]) - 1$ and $g(\mathbf{r}_1, \mathbf{r}_2; [\phi])$ is the usual pair distribution function (PDF) [21].

B.   Foundations of DFT

Density Functional Theory is based on a fundamental theorem first given by Mermin [5] stating that the grand potential of an electron gas in the presence of a one-body potential is a unique functional of the local density. The theorem is a finite temperature generalization of the Hohenberg–Kohn theorem which applies to zero-temperature systems and underlies the Density Functional Theory approach to *ab initio* quantum mechanical calculations. It is in these theorems that the two, now very different, disciplines called "Density Functional Theory" find their common roots. In keeping with the focus of this chapter, the proof given here will follow that found in Evans [13] and Hansen and McDonald [21] which are specific to a classical system.

The result is obtained in two steps: First, we form a functional over the space of distributions and show that the equilibrium distribution minimizes it. Second, this is used to show that two different external fields must give different local densities. This means there is a one-to-one relation between fields and densities so that the field can be eliminated in favor of the density, thus giving a functional of the density which is minimized by the equilibrium distribution.

Consider the space of distribution functions that are purely functionals of the applied external field as defined in Eq. (3). For brevity, these will here be denoted $f_N[\phi]$ with all other arguments being suppressed. Consider the functional

$$\Lambda[\phi; \phi_0] \equiv k_B T \sum_{N=0}^{\infty} \int (\ln(f_N[\phi]/f_N[\phi_0]) - \ln \Xi[\phi_0]) f_N[\phi] \, d\mathbf{p}^N d\mathbf{q}^N \qquad (9)$$

and note that $\Lambda[\phi_0; \phi_0] = -k_B T \ln \Xi[\phi_0]$, which is just $\Omega[\phi_0]$, the grand potential, so that

$$\Lambda[\phi; \phi_0] = \Lambda[\phi_0; \phi_0] + k_B T \sum_{N=0}^{\infty} \int f_N[\phi] \ln(f_N[\phi]/f_N[\phi_0]) \, d\mathbf{p}^N d\mathbf{q}^N \qquad (10)$$

Using the fact that $x \ln x \geq x - 1$ with equality only for $x = 1$, one has that

$$
\begin{aligned}
\sum_{N=0}^{\infty} \int & f_N[\phi] \ln(f_N[\phi]/f_N[\phi_0])\, d\mathbf{p}^N d\mathbf{q}^N \\
&= \sum_{N=0}^{\infty} \int f_N[\phi_0] \frac{f_N[\phi]}{f_N[\phi_0]} \ln(f_N[\phi]/f_N[\phi_0])\, d\mathbf{p}^N d\mathbf{q}^N \\
&\geq \sum_{N=0}^{\infty} \int f_N[\phi_0] \left(\frac{f_N[\phi]}{f_N[\phi_0]} - 1 \right) d\mathbf{p}^N d\mathbf{q}^N \\
&= 0
\end{aligned}
\tag{11}
$$

hence $\Lambda[\phi; \phi_0] \geq \Lambda[\phi_0; \phi_0]$ with equality only if the two functions are equal at all points. This completes the first step of the proof.

Next, suppose that $\phi \neq \phi_0$ and use the derived inequality and the explicit form of the distribution to find

$$
\begin{aligned}
\Lambda[\phi_0; \phi_0] < \Lambda[\phi; \phi_0] &= \sum_{N=0}^{\infty} \int (\hat{U}_N[\phi_0] - \hat{U}_N[\phi] - k_B T \ln \Xi[\phi]) f_N[\phi]\, d\mathbf{p}^N d\mathbf{q}^N \\
&= \int d\mathbf{r}\, (\phi_0(\mathbf{r}) - \phi(\mathbf{r})) \sum_{N=0}^{\infty} \int \hat{\rho}(\mathbf{r}) f_N[\phi]\, d\mathbf{p}^N d\mathbf{q}^N - k_B T \ln \Xi[\phi] \\
&= \Omega[\phi] + \int (\phi_0(\mathbf{r}) - \phi(\mathbf{r})) \rho(\mathbf{r}; [\phi])\, d\mathbf{r} \\
&= \Lambda[\phi; \phi] + \int (\phi_0(\mathbf{r}) - \phi(\mathbf{r})) \rho(\mathbf{r}; [\phi])\, d\mathbf{r}
\end{aligned}
\tag{12}
$$

Reversing the roles of ϕ and ϕ_0 gives

$$
\Lambda[\phi; \phi] < \Lambda[\phi_0; \phi_0] + \int (\phi(\mathbf{r}) - \phi_0(\mathbf{r})) \rho(\mathbf{r}; [\phi_0])\, d\mathbf{r}
\tag{13}
$$

If $\rho(\mathbf{r}; [\phi_0]) = \rho(\mathbf{r}; [\phi])$, then these two inequalities imply

$$
\Lambda[\phi_0; \phi_0] - \Lambda[\phi; \phi] < \int (\phi_0(\mathbf{r}) - \phi(\mathbf{r})) \rho(\mathbf{r}; [\phi])\, d\mathbf{r} < \Lambda[\phi_0; \phi_0] - \Lambda[\phi; \phi] \tag{14}
$$

which is a contradiction. Hence, different fields cannot generate the same average local density.

The conclusion is that there is a unique local density for a given external field and vice versa so that there is an invertible functional relation between average equilibrium density for a given field, $\rho(\mathbf{r}; [\phi]) \Leftrightarrow \phi(\mathbf{r}; [\rho])$. Since two equilibrium distributions, $f_N[\phi]$ and $f_N[\phi']$, differ only in the explicit form of the external field and its implicit effect on the partition function, the distribution is a functional of the field and, hence, of the local density. We can therefore interpret the functional Λ in this way by writing

$$
\Lambda[\phi; \phi_0] = \Lambda[\phi[\rho]; \phi_0] = \Omega[\rho; \phi_0]
\tag{15}
$$

where

$$\rho(\mathbf{r}; [\phi]) = \sum_{N=0}^{\infty} \int \hat{\rho}(\mathbf{r}) f(\mathbf{q}^N, \mathbf{p}^N, N; [\phi]) d\mathbf{q}^N d\mathbf{p}^N \qquad (16)$$

is the density corresponding to the field ϕ. Note that we use the notation Ω for this functional as well as for the grand partition function. This is justified by the fact that the functional chain rule implies that since $\Lambda[\phi; \phi_0]$ is minimized by the field $\phi = \phi_0$, so also $\Omega[\rho; \phi_0]$ is minimized by the equilibrium average density function, $\rho(\mathbf{r}) = \rho(\mathbf{r}; [\phi_0])$, corresponding to ϕ_0,

$$\left| \frac{\delta \Omega[\rho; \phi_0]}{\delta \rho} \right|_{\rho(\mathbf{r})=\rho(\mathbf{r};[\phi_0])} = 0 \qquad (17)$$

and $\Omega[\rho(\mathbf{r}; [\phi_0]); \phi_0] = \Omega$. Finally, it is convenient to separate out the part of $\Omega[\rho; \phi_0]$ which is independent of ϕ_0 and μ by defining

$$\Omega[\rho; \phi_0] = F[\rho] + \int \rho(\mathbf{r})(\phi_0(\mathbf{r})-\mu)d\mathbf{r} \qquad (18)$$

where

$$F[\rho] = \sum_{N=0}^{\infty} \int (\hat{K}_N + \hat{U}_N + k_B T \ln (N! h^{DN}) + k_B T \ln f[\phi[\rho]]) f[\phi[\rho]] \, d\mathbf{p}^N d\mathbf{q}^N \qquad (19)$$

has no explicit dependence on the field ϕ and, hence, *is unique for a given interaction potential*. This is the key conceptual point underlying classical density functional theory: The free energy functional $F[\rho]$ is independent of the applied field, and therefore the same model can be used for any external field. Once this functional is known, the average density for any given external field is calculated from the Euler–Lagrange equation

$$0 = \frac{\delta \Omega[\rho; \phi]}{\delta \rho(\mathbf{r})} \Rightarrow \frac{\delta F[\rho]}{\delta \rho(\mathbf{r})} = \mu - \phi(\mathbf{r}). \qquad (20)$$

This equation plays a central role in DFT: Given an external field and a model free energy functional, it can be solved to give the equilibrium density. Conversely, it defines the dependence of the field on the equilibrium density, $\phi(\mathbf{r}; \rho)$. Finally, given some independent determination of $\phi(\mathbf{r}; \rho)$, it can be integrated to give $F[\rho]$ as in the case of the ideal gas below.

Once the equilibrium profile is known, let us call it $\rho_0(\mathbf{r})$, the grand potential is given by $\Omega = \Omega[\rho_0; \phi]$. However, since ρ_0 is the solution to the Euler–Lagrange equation, the field dependence can be eliminated by noting that

$$\int \rho_0(\mathbf{r})(\mu - \phi(\mathbf{r}))d\mathbf{r} = \int \rho_0(\mathbf{r}) \left\{ \frac{\delta F[\rho]}{\delta \rho(\mathbf{r})} \right\}_{\rho(\mathbf{r})=\rho_0(\mathbf{r})} d\mathbf{r} \qquad (21)$$

If we abuse the notation somewhat and write

$$\left\{ \frac{\delta F[\rho]}{\delta \rho(\mathbf{r})} \right\}_{\rho(\mathbf{r})=\rho_0(\mathbf{r})} = \frac{\delta F[\rho_0]}{\delta \rho_0(\mathbf{r})} \tag{22}$$

then the grand potential becomes

$$\Omega = F[\rho_0] - \int \rho_0(\mathbf{r}) \frac{\delta F[\rho_0]}{\delta \rho_0(\mathbf{r})} \, d\mathbf{r} \tag{23}$$

These expressions should be interpreted with care because they are only valid when $\rho_0(\mathbf{r})$ is a solution of the Euler–Lagrange equation.

C. Integration of Functional Equations

There are two common circumstances in DFT in which the need arises to integrate functional differential equations. The first is the case of exactly solvable systems where one typicaly calculates the partition function for an arbitrary field, $\phi(\mathbf{r})$, differentiates with respect to the field to get the local density $\rho(\mathbf{r}; [\phi])$ and then inverts this relation to get the field as a functional of the density, $\phi(\mathbf{r}; [\rho])$. Then, integrating the Euler–Lagrange equation, Eq. (20), gives the key functional $F[\rho]$. The other circumstance is described in more detail later, but it has to do with the construction of approximate free energy functionals.

In either case, we recall that the equation

$$\frac{\delta F[\rho]}{\delta \rho(\mathbf{r}_1)} = c(\mathbf{r}_1; [\rho]) \tag{24}$$

where the right-hand side is some known functional, is integrable if and only if [22]

$$\frac{\delta c(\mathbf{r}_1; [\rho])}{\delta \rho(\mathbf{r}_2)} = \frac{\delta c(\mathbf{r}_2; [\rho])}{\delta \rho(\mathbf{r}_1)} \tag{25}$$

In this case, we can choose any two functions in density space, $\rho_0(\mathbf{r})$ and $\rho_1(\mathbf{r})$, and integrate between them along some path $\rho(\mathbf{r}; \lambda)$ where $\rho(\mathbf{r}; 0) = \rho_0(\mathbf{r})$ and $\rho(\mathbf{r}; 1) = \rho_1(\mathbf{r})$, giving

$$F[\rho_1] - F[\rho_0] = \int_0^1 d\lambda \int d\mathbf{r} \, \frac{\partial \rho(\mathbf{r}; \lambda)}{\partial \lambda} c(\mathbf{r}; [\rho]) \tag{26}$$

and the result is independent of the chosen path. In practice, the linear path $\rho(\mathbf{r}; \lambda) = \rho_0(\mathbf{r}) + \lambda(\rho_1(\mathbf{r}) - \rho_0(\mathbf{r}))$ is often used.

D. Expression for the Ideal Gas Contribution to the Free Energy Function

The final ingredient needed to make DFT useful is a way to relate the various functionals defined in the last subsection to physically meaningful quantities. In a system with no interactions, $\phi = 0$, one has

$$\Xi[\phi] = \exp(-\beta\Omega[\phi]) = \sum_{N=0}^{\infty} \frac{1}{N!h^{DN}} (2\pi k_B T)^{-DN/2} \left(\int \exp(-\beta(\phi(\mathbf{q})-\mu))d\mathbf{q} \right)^N$$

$$= \exp\left(\Lambda^{-D} \int \exp(-\beta(\phi(\mathbf{q})-\mu))d\mathbf{q} \right) \tag{27}$$

where $\Lambda = h/\sqrt{2m\pi k_B T}$ is the thermal wavelength. The average density is thus

$$\rho(\mathbf{r}; [\phi]) = \frac{\delta\Omega[\phi]}{\delta\phi(\mathbf{r})} = \Lambda^{-D}\exp(-\beta(\phi(\mathbf{r})-\mu)) \tag{28}$$

Denoting the ideal gas free energy functional by $F_{id}[\rho]$ and substituting into the Euler–Lagrange equation, Eq. (20), gives

$$\frac{\delta F_{id}[\rho]}{\delta\rho(\mathbf{r})} = \mu - \phi(\mathbf{r}; [\rho]) = k_B T \ln \Lambda^D \rho(\mathbf{r}) \tag{29}$$

which is integrated to give the explicit, exact result

$$\beta F_{id}[\rho] = \int_0^1 d\lambda \int d\mathbf{r}\, \rho(\mathbf{r}) \ln \Lambda^D \lambda\rho(\mathbf{r})$$

$$= \int (\rho(\mathbf{r}) \ln \Lambda^D \rho(\mathbf{r}) - \rho(\mathbf{r}))d\mathbf{r} \tag{30}$$

For interacting systems, the ideal-gas contribution is usually treated exactly, so one can write $F[\rho] = F_{id}[\rho] + F_{ex}[\rho]$ where the second term on the right is called the excess free energy functional. Thus, Euler–Lagrange equation becomes

$$k_B T \ln \rho(\mathbf{r}) + \frac{\delta F_{ex}[\rho]}{\delta\rho(\mathbf{r})} = \mu - \phi(\mathbf{r}) \tag{31}$$

A common method of solving this equation numerically is to write it in the form

$$\rho(\mathbf{r}) = \exp\left(\beta\mu - \beta\phi(\mathbf{r}) - \frac{\delta\beta F_{ex}[\rho]}{\delta\rho(\mathbf{r})} \right) \tag{32}$$

and to iterate by making an initial guess at $\rho(\mathbf{r})$, using this to evaluate the right-hand side and thus giving a new $\rho(\mathbf{r})$ and so forth. Mixing between the iterations can be useful.

E. A Simple Example: The Small Cavity

An exactly solvable model that plays an important role later is that of a hard sphere in a small cavity. Specifically, consider fields that are infinite outside some specified region, V, but are still considered to be arbitrary within V. If V is so small that the maximum number of hard spheres that can occupy it is one, then the partition function is just

$$\Xi[\phi] = \exp(-\beta\Omega[\phi]) = \sum_{N=0}^{\infty} \frac{1}{N! h^{DN}} (2\pi k_B T)^{-DN/2} \left(\int \exp(-\beta(\phi(\mathbf{q})-\mu)) d\mathbf{q} \right)^N$$

$$= 1 + \Lambda^{-D} \int_V \exp(-\beta(\phi(\mathbf{q}) - \mu)) d\mathbf{q} \tag{33}$$

Hence, the relation between the equilibrium density and the field is simply

$$\rho(\mathbf{r}) = -k_B T \frac{\delta \ln \Xi[\phi]}{\delta\phi(\mathbf{r})} = \frac{\Lambda^{-D}\exp(-\beta(\phi(\mathbf{r}) - \mu))}{1 + \Lambda^{-D}\int_V \exp(-\beta(\phi(\mathbf{q})-\mu)) d\mathbf{q}} \tag{34}$$

Integrating this and rearranging gives

$$\Lambda^{-D} \int_V \exp(-\beta(\phi(\mathbf{q}) - \mu)) d\mathbf{q} = \frac{\langle N \rangle}{1 - \langle N \rangle}, \qquad \langle N \rangle = \int_V \rho(\mathbf{r}) d\mathbf{q} \tag{35}$$

The Euler–Lagrange equation then becomes

$$\frac{\delta\beta F[\rho]}{\delta\rho(\mathbf{r})} = \beta(\mu - \phi(\mathbf{r})) = \ln\frac{\Lambda^D \rho(\mathbf{r})}{1 - \langle N \rangle} \tag{36}$$

so

$$\beta F[\rho] = \int_0^1 d\lambda \int d\mathbf{r}\, \rho(\mathbf{r}) \ln\frac{\Lambda^D \lambda\rho(\mathbf{r})}{1-\lambda\langle N \rangle}$$

$$= \int d\mathbf{r}\, \rho(\mathbf{r})\ln\Lambda^D\rho(\mathbf{r}) + (1-\langle N \rangle)(\ln(1-\langle N \rangle)) \tag{37}$$

$$= \beta F_{\mathrm{id}}[\rho] + (1-\langle N \rangle)(\ln(1-\langle N \rangle)) + \langle N \rangle$$

Note that this result is quite generally independent of the details of the size and shape of the cavity and only depends on the condition that the cavity cannot hold more than one hard sphere.

F. Exact General Expression for the Excess Part of the Free Energy Functional

Recall the relation

$$-\frac{\delta\rho(\mathbf{r}_1; [\phi])}{\delta\beta\phi(\mathbf{r}_2)} = \rho(\mathbf{r}_1)\delta(\mathbf{r}_1 - \mathbf{r}_2) + \rho(\mathbf{r}_1)\rho(\mathbf{r}_2)h(\mathbf{r}_1, \mathbf{r}_2) \tag{38}$$

derived above (see Eq. 8). Let us write the inverse relation as

$$\frac{\delta\beta\phi(\mathbf{r}_2; [\rho])}{\delta\rho(\mathbf{r}_1)} = -\frac{1}{\rho(\mathbf{r}_1)}\delta(\mathbf{r}_1 - \mathbf{r}_2) + \Gamma(\mathbf{r}_1, \mathbf{r}_2; [\rho]) \tag{39}$$

where we must clarify the nature of $\Gamma(\mathbf{r}_1, \mathbf{r}_2)$. Substituting these into the functional chain rule,

$$\delta(\mathbf{r}_1 - \mathbf{r}_3) = \int \frac{\delta\rho(\mathbf{r}_1; [\phi])}{\delta\beta\phi(\mathbf{r}_2)} \frac{\delta\beta\phi(\mathbf{r}_2; [\rho])}{\delta\rho(\mathbf{r}_3)} d\mathbf{r}_2$$

gives the relation

$$h(\mathbf{r}_1, \mathbf{r}_3) = \Gamma(\mathbf{r}_1, \mathbf{r}_3; [\rho]) + \int h(\mathbf{r}_1, \mathbf{r}_2)\rho(\mathbf{r}_2)\Gamma(\mathbf{r}_2, \mathbf{r}_3; [\rho])d\mathbf{r}_2 \qquad (40)$$

This is recognized as the Ornstein–Zernike equation for an inhomogeneous system [21] so that we can identify the unknown function Γ as the (two-body) direct correlation function (DCF), $\Gamma(\mathbf{r}_2, \mathbf{r}_3) = c_2(\mathbf{r}_2, \mathbf{r}_3; [\rho])$.

The excess part of the free energy can be understood by taking advantage of the fact that it is independent of the applied field. In particular, in the case of the equilibrium field, $\phi(\mathbf{r}) = \phi(\mathbf{r}; [\rho])$, a functional differentiation of the Euler–Lagrange equation, Eq. (20), gives

$$\frac{\delta^2\beta F_{ex}[\rho]}{\delta\rho(\mathbf{r}_1)\delta\rho(\mathbf{r}_2)} = -\frac{1}{\rho(\mathbf{r}_1)}\delta(\mathbf{r}_1 - \mathbf{r}_2) - \frac{\delta\beta\phi(\mathbf{r}_1; [\rho])}{\delta\rho(\mathbf{r}_2)}$$
$$= -c_2(\mathbf{r}_1, \mathbf{r}_2; [\rho]) \qquad (41)$$

This result is very important and deserves to be highlighted: The functional $F_{ex}[\rho]$ is a general functional of its argument, ρ, independent of the applied field. We have derived its form by assuming a particular field—namely, the equilibrium field $\phi[\rho]$. However, having determined it for this field, it is the same for all other fields as well. This therefore completes the specification of the general functional $\Omega[\rho; \phi_0]$.

This result can be used to give an expression for the exact excess free energy functional. Since Eq. (41) is an exact relation and since the two-body DCF is an exact derivative, it can be integrated through density space to give an exact relation between the excess free energy functional for the density $\rho_0(\mathbf{r})$ and that for the density, $\rho_1(\mathbf{r})$. If one forms a path between these two density profiles parameterized by some scalar such as $\rho_\lambda(\mathbf{r}) = (1 - \lambda)\rho_0(\mathbf{r}) + \lambda\rho_1(\mathbf{r})$, then the result is

$$\beta F_{ex}[\rho_1] = \beta F_{ex}[\rho_0] + \int_0^1 d\lambda \int d\mathbf{r}_1 \left[\frac{\delta\beta F_{ex}[\rho_\lambda(\mathbf{r}_1)]}{\delta\rho_\lambda(\mathbf{r}_1)} \right]_{\rho_0} \frac{\partial\rho_\lambda(\mathbf{r}_1)}{\partial\lambda}$$
$$- \int_0^1 d\lambda \int_0^1 d\lambda' \int d\mathbf{r}_1 d\mathbf{r}_2 c_2(\mathbf{r}_1, \mathbf{r}_2; [\rho_{\lambda'}]) \frac{\partial\rho_{\lambda'}(\mathbf{r}_1)}{\partial\lambda'} \frac{\partial\rho_{\lambda'}(\mathbf{r}_2)}{\partial\lambda'} \qquad (42)$$

Note that this is independent of the parameterization chosen. From the equivalence of fields and densities, there will be some field that generates the density profile $\rho_0(\mathbf{r})$ at the given chemical potential. Calling this field $\phi(\mathbf{r}_1; [\rho_0])$, the

Euler–Lagrange equation can be used, giving

$$\beta F_{ex}[\rho_1] = \beta F_{ex}[\rho_0] + \int d\mathbf{r}_1 [\beta\mu - \ln \rho_0(\mathbf{r}_1) - \beta\phi(\mathbf{r}_1; [\rho_0])](\rho_1(\mathbf{r}_1) - \rho_0(\mathbf{r}_1))$$

$$- \int_0^1 d\lambda \int_0^\lambda d\lambda' \int d\mathbf{r}_1 d\mathbf{r}_2 c_2(\mathbf{r}_1, \mathbf{r}_2; [\rho_{\lambda'}]) \frac{\partial \rho_{\lambda'}(\mathbf{r}_1)}{\partial \lambda'} \frac{\partial \rho_{\lambda'}(\mathbf{r}_2)}{\partial \lambda'}$$

$$(43)$$

Specializing to the case that the reference state is a liquid, $\rho_0(\mathbf{r}) = \bar{\rho}_0$, the field $\beta\phi(\mathbf{r}_1; [\rho_0])$ will be a constant such that $\mu - \phi(\mathbf{r}_1; [\rho_0])$ will be chemical potential that generates $\bar{\rho}_0$ which, though an abuse of notation, we will denote as $\mu(\bar{\rho}_0)$. This should not be confused with the applied chemical potential μ, which is an external parameter in the grand canonical ensemble. Then, taking the linear parameterization through density space, the excess functional is

$$\frac{1}{V}\beta F_{ex}[\rho_1] = \beta f_{ex}(\bar{\rho}_0) + \frac{\partial \beta f_{ex}(\bar{\rho}_0)}{\partial \bar{\rho}_0}(\bar{\rho}_1 - \bar{\rho}_0)$$

$$- \frac{1}{V}\int_0^1 d\lambda \int_0^\lambda d\lambda' \int d\mathbf{r}_1 d\mathbf{r}_2 c_2(\mathbf{r}_1, \mathbf{r}_2; [(1-\lambda')\rho_0 + \lambda'\rho_1])$$

$$\times (\rho_1(r_1) - \bar{\rho}_0)(\rho_1(\mathbf{r}_2) - \bar{\rho}_0) \quad (44)$$

This is still not useful because it involves the unknown two-body direct correlation function for an arbitrary density distribution. The idea underlying many DFT models is that, at least in the liquid state, the DCF is relatively simple in structure compared to other properties such as the pair distribution function. For example, the PDF in a dense simple fluid has a slowly decaying oscillatory structure describing successive shells of neighbors, whereas, for example, in the hard-sphere fluid, the DCF is well-approximated as a monotonic cubic function vanishing outside the hard core. Hence, it was hoped that relatively crude approximations to the DCF might be adequate.

One way to implement this intuition is to perform a functional Taylor expansion of the DCF about another reference liquid density,

$$c_2(\mathbf{r}_1, \mathbf{r}_2; [\rho_\lambda]) = c_2(\mathbf{r}_{12}; \bar{\rho}(\lambda))$$

$$+ \sum_{n=3}^\infty \frac{1}{(n-2)!} \int d\mathbf{r}_3 \ldots d\mathbf{r}_N c_n(\mathbf{r}_3, \ldots, \mathbf{r}_N; \bar{\rho}(\lambda))$$

$$(\rho_\lambda(\mathbf{r}_3) - \bar{\rho}(\lambda)) \ldots (\rho_\lambda(\mathbf{r}_N) - \bar{\rho}(\lambda)) \quad (45)$$

where the right-hand side is written using the higher-order direct correlation functions for the liquid,

$$c_N(\mathbf{r}_3, \ldots, \mathbf{r}_N; \bar{\rho}) = \frac{\delta^{N-2} c_2(\mathbf{r}_1, \mathbf{r}_2; [\rho])}{\delta\rho(\mathbf{r}_3) \ldots \delta\rho(\mathbf{r}_N)}\bigg|_{\rho(\mathbf{r}) = \bar{\rho}} \quad (46)$$

Equations (44) and (45) give an exact expression of the excess free energy functional in terms of the properties of a uniform fluid. It is independent of the path taken through density space so that one can choose, for example, $\rho_\lambda(\mathbf{r}) = \lambda \rho(\mathbf{r})$ and it is also exact for all choices of $\bar{\rho}_0$ and $\bar{\rho}(\lambda)$. In principle, this gives a method to describe *nonuniform* systems, even solids, based only on knowledge of the uniform fluid. Unfortunately, little is known about the higher-order direct correlation functions so that in practice, only the first three terms of the expansion are used. A simple approximation consists of truncation of the higher-order terms and the choice $\bar{\rho}(\lambda) = \bar{\rho}_0$, giving

$$\frac{1}{V}\beta F_{\text{ex}}[\rho_1] \simeq \beta f_{\text{ex}}(\bar{\rho}_0) + \frac{\partial \beta f_{\text{ex}}(\bar{\rho}_0)}{\partial \rho_0}(\bar{\rho}_1 - \bar{\rho}_0)$$

$$-\frac{1}{2V}\int d\mathbf{r}_1 d\mathbf{r}_2 c_2(r_{12}; \bar{\rho}_0)(\rho_1(\mathbf{r}_1) - \bar{\rho}_0)(\rho_1(\mathbf{r}_2) - \bar{\rho}_0) \qquad (47)$$

However, this is not really satisfactory because it is inconsistent when the target system is itself the uniform fluid, that is, $\rho_1(\mathbf{r}) = \bar{\rho}_1$. To remedy this, one should take $\bar{\rho}_0 = \bar{\rho}_1$, which is the oldest and simplest approximate DFT, first studied by Ramakrishnan and Yussouff [6]:

$$\frac{1}{V}\beta F_{\text{ex}}[\rho_1] - \frac{1}{V}\beta F_{\text{ex}}(\bar{\rho}_1) \simeq -\frac{1}{2V}\int d\mathbf{r}_1 d\mathbf{r}_2 c_2(\mathbf{r}_{12}; \bar{\rho}_0)(\rho_1(\mathbf{r}_1) - \bar{\rho}_1)(\rho_1(\mathbf{r}_2) + \bar{\rho}_1)$$

$$(48)$$

This approximation, which involves only knowledge of the DCF in the liquid, is still a standard starting point for calculations in which simplicity is favored over accuracy.

G. Correlation Functions

Given the free energy functional $F[\rho]$, the entire hierarchy of direct correlation functions follows immediately by functional differentiation. In many cases, however, it is more useful to have the pair distribution function, $g(\mathbf{r}_1, \mathbf{r}_2; \mu, [\phi])$, giving the probability to find a particle at position $\mathbf{r}_2$ given that their is one at position $\mathbf{r}_1$. One method is to use the Ornstein–Zernike equation for inhomogeneous fluids [see Eq. (40)]. However, as pointed out by Percus [23, 24], DFT provides another method of obtaining the PDF which can be easier to implement in practice. Suppose the system interacts via a two-body potential, $v(\mathbf{r}_1, \mathbf{r}_2)$, and is subject to an external potential $\phi(\mathbf{r})$. The two-body distribution $\rho(\mathbf{r}_1, \mathbf{r}_2; \mu, [\phi])$ is the probability to find one particle at position $\mathbf{r}_1$ and another at position $\mathbf{r}_2$. It is related to the PDF by $\rho(\mathbf{r}_1, \mathbf{r}_2; \mu, [\phi]) = \rho(\mathbf{r}_1; \mu, [\phi]) \times \rho(\mathbf{r}_2; \mu, [\phi])g(\mathbf{r}_1, \mathbf{r}_2; \mu, [\phi])$. Since the one-body density is the probability to find a particle at a given position, it follows that $\rho(\mathbf{r}_1, \mathbf{r}_2; \mu, [\phi])/\rho(\mathbf{r}_1; \mu, [\phi]) = \rho(\mathbf{r}_2; \mu, [\phi])g(\mathbf{r}_1, \mathbf{r}_2; \mu, [\phi])$ is the conditional distribution giving the probability to find a particle at position $\mathbf{r}_2$ given that there is one at position $\mathbf{r}_1$. Conceptually, this is identical to the probability to find a particle at position $\mathbf{r}_2$ in a system with a particle fixed at position $\mathbf{r}_1$. A fixed particle is equivalent to an external field acting on the rest of the system, so another way to interpret this is that $\rho(\mathbf{r}_2; \mu, [\phi]) \times g(\mathbf{r}_1, \mathbf{r}_2; \mu, [\phi])$ is the same as the density profile in a system with the same external field *plus* the field $v(\mathbf{r}_1, \mathbf{r}_2)$,

$$\rho(\mathbf{r}_2; \mu, [\phi]) g(\mathbf{r}_1, \mathbf{r}_2; \mu, [\phi]) = \rho(\mathbf{r}_2; \mu, [\phi'_{r_1}]) \tag{49}$$

where

$$\phi'_{r_1} = \phi'(\mathbf{r}) + v(\mathbf{r}_1, \mathbf{r}). \tag{50}$$

Using the Euler–Lagrange equation, Eq. (32), this can be written as

$$\begin{aligned}
&\rho(\mathbf{r}_2; \mu, [\phi]) g(\mathbf{r}_1, \mathbf{r}_2; \mu, [\phi]) \\
&= \exp\left(\beta\mu - \beta\phi(\mathbf{r}_2) - \beta v(\mathbf{r}_1, \mathbf{r}_2) - \left. \frac{\delta\beta F_{\text{ex}}[n]}{\delta n(\mathbf{r}_2)} \right|_{n(\mathbf{r}) = \rho(\mathbf{r}; \mu, [\phi'_{r_1}])} \right)
\end{aligned} \tag{51}$$

Thus, by solving the DFT with the field ϕ'_{r_1}, one obtains the PDF for the system with field ϕ. In particular, if $\phi = 0$, one finds that the PDF of the bulk fluid is given by

$$\bar{\rho}(\mu) g(\mathbf{r}_1, \mathbf{r}_2; \mu) = \exp\left(\beta\mu - \beta v(\mathbf{r}_1, \mathbf{r}_2) - \left. \frac{\delta\beta F_{\text{ex}}[n]}{\delta n(\mathbf{r}_2)} \right|_{n(\mathbf{r}) = \rho(\mathbf{r}; \mu, [\phi'_{r_1}])} \right) \tag{52}$$

These relations are often used in practical calculations to get the PDF from a DFT calculation. Furthermore, since the same DFT allows one to calculate both the PDF and the DCF and since they are related via the Ornstein–Zernike relation, this gives a method to check the self-consistency of model calculations analogous to the comparison of the virial and compressibility routes to the equation of state in liquid-state theory.

H. Parameterized Profiles and Gradient Expansions

In a typical application of DFT, given some approximation to the free energy functional, the Euler–Lagrange equations are solved to get the equilibrium density profile and, from this, the free energy is calculated. Since the density profile is a function, this procedure involves discretization and can become computationally very expensive. In many cases, one is able to make a reasonable guess as to the general properties of the density profile and can therefore propose an analytic form that is expected to closely approximate the exact result. For example, in the case of a planar interface in which the density is uniform except in one direction, a hyperbolic tangent is a natural choice:

$$\rho(z; \bar{\rho}_{-\infty}, \bar{\rho}_{\infty}, z_0, a) = \bar{\rho}_{-\infty} + (\bar{\rho}_{\infty} - \bar{\rho}_{-\infty}) \frac{\exp(a(z - z_0))}{\exp(a(z - z_0)) + \exp(-a(z - z_0))}$$
(53)

so that there are four parameters: the densities at $z = \pm\infty$, the location of the interface, z_0, and the inverse width, a. A similar form, with the Cartesian coordinate z replaced by the radial coordinate r, might be used to described a spherical cluster (droplet or bubble in a liquid–vapor system). A very important parameterization used in many calculations of solids is to approximate the density as a sum of Gaussians centered at the lattice sites:

$$\rho(\mathbf{r}; \alpha, x, \bar{\rho}_{\text{latt}}) = x \sum_{n=0}^{\infty} \left(\frac{\alpha}{\pi}\right)^{3/2} \exp(-\alpha(\mathbf{r} - \mathbf{R}_n)^2)$$
(54)

where the sum is over lattice vectors, $\mathbf{R}_n$, the magnitudes of which depend on the lattice density $\bar{\rho}_{\text{latt}}$, where α controls the width of the Gaussians and $0 < x \leq 1$ is the occupancy that allows for the possibility that not all lattice sites are occupied. This is actually a very flexible parameterization, as can be seen when it is written in Fourier space as

$$\rho(\mathbf{r}; \alpha, x, \bar{\rho}_{\text{latt}}) = x\bar{\rho}_{\text{latt}} \sum_{n=0}^{\infty} \exp(i\mathbf{K}_n \cdot \mathbf{r}) \exp(-K_n^2/4\alpha)$$
(55)

where the sum is now over all reciprocal lattice vectors, $\mathbf{K}_n$. In this form, it is clear that $\lim_{\alpha \to 0} \rho(\mathbf{r}; \alpha, x, \bar{\rho}_{\text{latt}}) = x\bar{\rho}_{\text{latt}}$, which is the uniform fluid limit. Hence, the Gaussian approximation can be used to approximate both liquid-like and solid-like systems and, most importantly, the transition from one to another. Note that with this parameterization, the calculation of the ideal contribution to the free energy is not trivial. If the lattice parameter is denoted a, then for $\alpha a^2 \ll 1$, asymptotic expressions are available [25], but these are not very useful. For $\alpha a^2 \gg 1$, a simple calculation [25] gives $F_{\text{id}}[\rho] \simeq \frac{3}{2} \ln(\alpha\Lambda^2/\pi) - \frac{5}{2}$, which becomes essentially exact for $\alpha a^2 > 100$. At intermediate values, the calculation must be performed numerically. The accuracy of the Gaussian parameterization has been the checked in several studies, and it seems to always be a very good first approximation [26–28].

Denote an arbitrary parameterized profile as $\rho(\mathbf{r}; \Gamma)$, where $\Gamma = \{\Gamma_i\}_{i=1}^{n}$ represents a set of n parameters. If there is a set of parameters so that $\rho(\mathbf{r}; \Gamma)$ is the exact equilibrium functional, then

$$\frac{\partial \Omega[\rho; \phi_0]}{\partial \Gamma_i} = \int \frac{\delta \Omega[\rho; \phi_0]}{\delta \rho(\mathbf{r}; \Gamma)} \frac{\partial \rho(\mathbf{r}; \Gamma)}{\partial \Gamma_i} d\mathbf{r} = 0$$
(56)

so

$$\frac{\partial F[\rho]}{\partial \Gamma_i} = \int \frac{\partial \rho(\mathbf{r}; \Gamma)}{\partial \Gamma_i} (\mu - \phi(\mathbf{r})) d\mathbf{r} \tag{57}$$

If the only effect of the field is to confine the system to a volume V, then this becomes

$$\frac{\partial F[\rho]}{\partial \Gamma_i} = \mu \frac{\partial \bar{\rho}(\Gamma)}{\partial \Gamma_i} V, \qquad \bar{\rho}(\Gamma) V \equiv \int \rho(\mathbf{r}; \Gamma) d\mathbf{r} \tag{58}$$

which is the key equation that serves to fix the parameters in practical calculations.

In some cases, a simple parameterization of this form is insufficient. For example, to describe a liquid–solid interface, one would might use the Gaussian parameterization but with values of the average density and of α that vary as one moves from the liquid region to the solid region. There is no difficulty in extending the discussion of parameterized profiles to this case, but even with the parameterization the calculations can be quite expensive. If the variation of the parameters is expected to be slow relative to the atomic and interfacial length scales, then one might imagine performing a gradient expansion of the free energy. There are actually two versions of the gradient expansion in use [29] and both are based on a parameterization of the density appropriate for liquid–solid interfaces. The first is due to Evans [13] and refined by Oxtoby and Haymet [30, 31]. Imagine that we have some parameterization, $\rho(\mathbf{r}; \Gamma(\mathbf{r}))$, and calculate the (Helmholtz) free energy difference between this system and that of a uniform fluid at second order in perturbation theory [see Eq. (48)],

$$\beta F[\rho] - \beta F(\bar{\rho}_0) \simeq \beta F_{id}[\rho] - \beta F_{id}(\bar{\rho}_0)$$
$$- \frac{1}{2V} \int d\mathbf{r}_1 d\mathbf{r}_2 c_2(r_{12}; \bar{\rho}_0)(\rho(\mathbf{r}_1; \Gamma(\mathbf{r}_1)) - \bar{\rho})(\rho(\mathbf{r}_2; \Gamma(\mathbf{r}_2)) - \bar{\rho})$$
$$= \beta \int \Delta f(\rho(\mathbf{r}; \Gamma(\mathbf{r})); \bar{\rho}_0) d\mathbf{r}_1 - \frac{1}{2V} \int d\mathbf{r}_1 d\mathbf{r}_2 c_2(r_{12}; \bar{\rho}_0)$$
$$\times (\rho(\mathbf{r}_1; \Gamma(\mathbf{r}_1)) - \bar{\rho})(\rho(\mathbf{r}_2; \Gamma(\mathbf{r}_2)) - \rho(\mathbf{r}_2; \Gamma(\mathbf{r}_1))) \tag{59}$$

where the second line uses the uniform free energy difference per unit volume,

$$\Delta f(\rho; \bar{\rho}_0) = \frac{1}{V} F(\rho) - \frac{1}{V} F(\bar{\rho}_0) \tag{60}$$

Thus, the first term looks like a local free energy contribution while the second depends on $\rho_1(\mathbf{r}_2; \Gamma(\mathbf{r}_2)) - \rho_1(\mathbf{r}_2; \Gamma(\mathbf{r}_1))$, which can then be expanded in gradients of $\Gamma(\mathbf{r})$. Further development of this model depends on an explicit choice for the parameterization,

$$\rho(\mathbf{r}; \Gamma(\mathbf{r})) = \rho_0 \left(1 + \Gamma_0(\mathbf{r}) + \sum_{n=1} \exp(i\mathbf{K}_n \cdot \mathbf{r})\Gamma_n(\mathbf{r}) \right) \tag{61}$$

The result, truncated at second order in the gradient expansion, is

$$\beta F[\rho] - \beta F(\bar{\rho}_0) \simeq \beta \int \left\{ \Delta f(\rho(r; \Gamma(\mathbf{r})); \bar{\rho}_0) + \frac{1}{2} \sum_{i,j=1}^{D} \sum_{a,b=1}^{N} K_{ij}^{ab} \frac{\partial \Gamma_a(\mathbf{r})}{\partial r_i} \frac{\partial \Gamma_b(\mathbf{r})}{\partial r_j} \right\} d\mathbf{r} \tag{62}$$

with

$$K_{ij}^{ab} = \delta_{ab} \frac{1}{2} \rho_0^2 \int \exp(i\mathbf{K}_a \cdot \mathbf{r}) r_i r_j c_2(r; \bar{\rho}_0) d\mathbf{r} \tag{63}$$

and can be found in the cited papers. One feature of this model is that the local free energy term still involves $\rho(\mathbf{r}; \Gamma(\mathbf{r}))$, so that the long-wavelength and short-wavelength variations of the density are not completely separated.

An alternative expansion was given by Löwen, Beyer, and Wagner [32, 33] and further developed by Lutsko [29]. The idea is that space is divided into Wigner–Seitz cells centered on the lattice sites. Within each cell, the free energy is functionally Taylor-expanded about the value of the density parameters at the center of the cell; and the result, which will depend on terms like $\Gamma(\mathbf{r}) - \Gamma(\mathbf{R}_n)$, is again expanded in gradients of the parameters. It turns out that if the gradient expansion is truncated at second order, this automatically truncates the functional expansion at second order as well. At this point, the free energy has the gradient form, but is written as a sum over Wigner–Seitz cells. The transition to a continuum description is subtle and requires the further neglect of contributions of higher order in the gradients of the density parameters. The final result is

$$\beta F[\rho] \simeq \beta \int \left\{ \frac{1}{V} \beta \tilde{F}(\Gamma(\mathbf{r})) + \frac{1}{2} \sum_{i,j=1}^{D} \sum_{a,b=1}^{N} K_{ij}^{ab}(\Gamma(\mathbf{r})) \frac{\partial \Gamma_a(\mathbf{r})}{\partial r_i} \frac{\partial \Gamma_b(\mathbf{r})}{\partial r_j} \right\} d\mathbf{r} \tag{64}$$

where

$$\tilde{F}(\Gamma) = \beta F[\rho_\Gamma]$$
$$K_{ij}^{ab}(\Gamma) = \frac{1}{2V} \int r_{12,i} r_{12,j} c_2(\mathbf{r}_1, \mathbf{r}_2; [\rho_\Gamma]) \frac{\partial \rho(\mathbf{r}_1; \Gamma)}{\partial \Gamma_a} \frac{\partial \rho(\mathbf{r}_2; \Gamma)}{\partial \Gamma_b} d\mathbf{r}_1 d\mathbf{r}_2 \tag{65}$$

and the notation $F[\rho_\Gamma]$ indicates the functional $F[\rho]$ evaluated with the function $\rho(\mathbf{r}; \Gamma)$—that is, for fixed values of Γ. This result is more general and more complex than the previous one. The DCF occurring here is that for the system with uniform density parameters, not the uniform fluid as in the previous case. It also does not depend on any particular parameterization of the density. In fact, if one makes the approximation $c_2(\mathbf{r}_1, \mathbf{r}_2; [\rho_\Gamma]) \sim c_2(r_{12}; \bar{\rho}_0)$ and uses the Haymet–Oxtoby parameterization for the density, then the coefficients $K_{ij}^{ab}(\Gamma)$ become the same in both theories.

One important aspect of both theories is that the expression for the coefficient K_{ij}^{ab} will only be finite if the DCF is short-ranged (or goes to zero sufficiently quickly so that the second moment exists). In fact, the derivations implicitly assume that analogous higher-order terms also exist so that in general the interactions should be short-ranged. This is not a problem if the potential is truncated, as is typically the case when comparing to simulation, or if the free energy is separated into a short-ranged and a long-ranged contribution, as is discussed in more detail below, and the gradient expansion only applied to the short-ranged part. A final comment is that the use of gradient expansions has been questioned in general on the grounds that they treat correlations too crudely [34]. This criticism seems more apropos of the first form of the gradient model where one is using a very crude model for the DCF function (namely that of a uniform liquid). In the more sophisticated model, no such assumption is made and, in fact, the truncation of the functional Taylor expansion at second order is an exact consequence of the truncation of the gradient expansion at second order, thus suggesting that this criticism may carry less weight.

III. DFT MODELS BASED ON THE LIQUID STATE

A. A Preview of DFT: The Square Gradient Model

The previous section dealt with the theoretical basis for DFT without giving much indication as to how it could be used in practice. Toward this end, we observe that some knowledge is available for the liquid state. If there is no applied field, then there is no source of spatial anisotropy and the average density must be a constant, $\rho(\mathbf{r}) = \bar{\rho}$. (Note that we often speak of solids in the absence of fields and these are not translationally invariant; however, there must actually be a field that serves to fix the position and orientation of the crystal lattice: Averaging over all translations and rotations of the lattice would give a uniform density. We assume it is enough that such a field act at the boundaries of the system and that its effect is negligible in the thermodynamic limit.) Rather than having no field at all, we will assume the system is in a container, which means that the field is zero in the interior region and infinite outside this region. Again, we assume that the interior can be made so large that the surface effects of the container are negligible. Then, if the available volume is V, the grand potential is $\frac{1}{V}\Omega = \Lambda[\bar{\rho}; \phi[\bar{\rho}] = 0] = \bar{\rho} \ln \Lambda^D \bar{\rho} - \bar{\rho} + \frac{1}{V} F_{\text{ex}}[\bar{\rho}] - \mu \bar{\rho}$; hence, it follows that

$\frac{1}{V} F_{ex}[\bar{\rho}]$ is the excess Helmholtz free energy for a uniform liquid which will be denoted as $f_{ex}(\bar{\rho})$. Whenever necessary, it is assumed in DFT that this function as well as other properties of the uniform liquid such as its two-body DCF, $c_2(\mathbf{r}_1, \mathbf{r}_2; [\bar{\rho}]) = c_2(r_{12}; \bar{\rho})$, the pair distribution function, and so on, are known or knowable from liquid-state theory (thermodynamic perturbation theory or integral equation theories, see, for example, Ref. 21).

It might seem that one could make a simple DFT model based on the (known) free energy function of the liquid by postulating that

$$F_{ex}[\rho] \simeq \int f_{ex}(\rho(\mathbf{r})) d\mathbf{r} \qquad (66)$$

which is known as the "local density approximation." However, a moment's thought shows that this is too simple. For example, in the case of a planar liquid–vapor interface, the grand potential of the liquid and vapor are the same (by definition of coexistence). Since the coexisting phases are minima of the grand potential, the free energy for densities between that of the liquid and vapor must be larger, thus implying that the density functional is minimized by a system with an infinitely thin interface, which is a very crude and not very useful approximation. This defect can be corrected, as first discussed by van der Waals himself [1, 2], by realizing that gradients of the density must be energetically costly since a small volume in the system with neighboring volumes having different densities will necessarily feel a net force. Calculating the forces in the limit of slowly varying densities leads to the well-known "square-gradient approximation" [1, 2],

$$F_{ex}[\rho] \simeq \int \left[f_{ex}(\rho(\mathbf{r})) + \frac{1}{2} g (\nabla\rho(\mathbf{r}))^2 \right] d\mathbf{r} \qquad (67)$$

The parameter g is, at this point, unknown and is often treated phenomenologically. As discussed above, it can be calculated from knowledge of the DCF. To give a flavor of the types of calculations performed using DFT, we can use this model to calculate the structure of the liquid–vapor interface. Let the densities of the coexisting liquid and vapor at some given temperature and chemical potential be $\bar{\rho}_l$ and $\bar{\rho}_\phi$, respectively. Then, for a planar interface we expect the density to depend only on one coordinate, say the z-coordinate, and to take the values $\rho(\infty) = \bar{\rho}_l$ and $\rho(-\infty) = \bar{\rho}_v$. The Euler–Lagrange equation becomes

$$\frac{df(\rho(z))}{d\rho(z)} - g \frac{d^2\rho(z)}{dz^2} = \mu \qquad (68)$$

where $f(\rho) = \rho \ln \rho - \rho + f_{ex}(\rho)$. Multiplying through by $d\rho(z)/dz$ and integrating under the assumption that the derivatives vanish at large $|z|$ gives

$$f(\rho(z)) - \frac{1}{2} g \left(\frac{d\rho(z)}{dz} \right)^2 - \mu\rho(z) = f(\bar{\rho}_v) - \mu\bar{\rho}_v \equiv \omega(\bar{\rho}_v; \mu) \qquad (69)$$

This equation can be solved by quadratures and, given an analytic form of $f_{ex}(\rho)$, the profile determined numerically. The excess free energy per unit area, the surface free energy γ which is often assumed to be the same as the surface tension, can then be calculated from

$$\gamma = \int_{-\infty}^{\infty} \left(f(\rho(z)) + \frac{1}{2}g\left(\frac{d\rho(z)}{dz}\right)^2 - \mu\rho(z) - \omega(\bar{\rho}_v; \mu) \right) dz \qquad (70)$$

If the profile is monotonic, one can solve Eq. (69) for $d\rho/dz$ and write this as

$$\gamma = \sqrt{2g} \int_{\bar{\rho}_v}^{\bar{\rho}_l} \sqrt{f(\rho) - \mu\rho - \omega(\bar{\rho}_v; \mu)} \, dn \qquad (71)$$

so that the surface free energy can be determined by a simple integration without even solving for the profile.

B.   A Survey of Models

One of the applications that motivated the development of early DFT models was the description of freezing and particularly the description of hard-sphere freezing. It was always recognized that if good models could be created for hard-spheres, then realistic potentials with long-ranged attractive interactions could, at worst, be treated perturbatively (an assumption justified by the extension of thermodynamic perturbation theory to the solid state [35]). At best, a successful model for hard-spheres would be immediately generalizable to simple fluids and beyond. The strategy that dominated much of the early work was to try to somehow use information about the uniform liquid to construct a functional applicable to nonuniform systems, particularly the solid that in this context is viewed as a highly nonuniform liquid. In the remainder of this section, a variety of models based on this approach is reviewed and their successes and shortcomings are pointed out. A more thorough discussion of these developments can be found in Evans [13].

1.   Models Based on Perturbation Theory about the Uniform Liquid State

a.   The   Ramakrishnan–Yussouff/Haymet–Oxtoby   Theory.   Perhaps   the   earliest successful DFT of freezing was that due to Ramakrishnan and Yussouff [6] and further developed by Haymet and Oxtoby [30]. It is based on the truncation of the perturbative expansion, Eqs. (44) and (45), with the simplest choices $\bar{\rho}_0 = \bar{\rho}(\lambda) = \bar{\rho}_1$ giving

$$\frac{1}{V}\beta F_{ex}[\rho_1] = \beta f_{ex}(\bar{\rho}_0) - \frac{1}{2V}\int d\mathbf{r}_1 d\mathbf{r}_2 c_2(r_{12}; \bar{\rho}_0)(\rho_1(\mathbf{r}_1) - \bar{\rho}_0)(\rho_1(\mathbf{r}_2) - \bar{\rho}_0) + \cdots \quad (72)$$

Higher-order terms are not written as practical calculations are usually, but not always, done after truncating the expansion at second order. The suspicion that higher-order terms were not negligable, particularly the work of Curtin [36] indicating that the series converges slowly, if at all, led to a desire for "nonperturbative" alternatives to this simple theory. Nevertheless, it is still frequently used as a simplest first approximation.

b. The Effective Liquid Theory. Baus and Colot [37] proposed what was termed a "nonperturbative" theory, meaning it was not based on perturbation about a liquid with the same density as the inhomogeneous system (usually a solid, in this context). It can, however, still be viewed as a perturbative theory based on Eqs. (44, 45) with $\bar{\rho}_0 = \bar{\rho}_1$. The idea was that the unknown DCF be approximated by that of the liquid at a density, $\bar{\rho}_{ELA}$ for which the first peak of the structure factor of the liquid occurs at the smallest reciprocal lattice vector of the solid (thus, in some sense matching the structure of the liquid and solid). This is equivalent to taking $\bar{\rho}(\lambda) = \bar{\rho}_{ELA}$ in Eqs. (44)–(45) giving

$$\frac{1}{V}\beta F_{ex}[\rho_1] = \beta f_{ex}(\bar{\rho}_1) - \frac{1}{2V} \int d\mathbf{r}_1 d\mathbf{r}_2 c_2(r_{12}; \bar{\rho}_{ELA})(\rho_1(\mathbf{r}_1) - \bar{\rho}_1)(\rho_1(\mathbf{r}_2) - \bar{\rho}_1) + \cdots$$

(73)

which is the ELA result.

c. The Self-Consistent Effective Liquid Theory. Proposed by Baus [38], this theory involved an actual integration though density space. First, the initial reference is taken to be zero, $\bar{\rho}_0 = 0$, and then the choice $\bar{\rho}(\lambda) = \lambda \bar{\rho}_{SCELA}$ is made giving

$$\frac{1}{V}\beta F_{ex}[\rho_1] = -\frac{1}{V} \int_0^1 d\lambda \int_0^\lambda d\lambda' \int d\mathbf{r}_1 d\mathbf{r}_2 c_2(r_{12}; \lambda' \bar{\rho}_{SCELA}) \rho_1(\mathbf{r}_1) \rho_1(r_2) + \cdots$$

$$= -\frac{1}{V} \int_0^1 d\lambda \int d\mathbf{r}_1 d\mathbf{r}_2 (1 - \lambda) c_2(r_{12}; \lambda \bar{\rho}_{SCELA}) \rho_1(\mathbf{r}_1) \rho_1(\mathbf{r}_2) + \cdots$$

(74)

where the second line follows from an integration by parts. The density $\bar{\rho}_{SCELA}$ is chosen to satisfy the self-consistency requirement that the excess free energy per atom be the same in the solid as in the reference system,

$$\psi_{ex}(\bar{\rho}_{SCELA}) \equiv \frac{1}{\bar{\rho}_{SCELA}} f_{ex}(\bar{\rho}_{SCELA}) = \frac{1}{\rho_1 V} F_{ex}[\rho_1]$$

(75)

The idea is that there is always some liquid density for which this equation—the thermodynamic mapping—holds true. Similarly, there is always some liquid density for which Eq. (74) holds true, the so-called structural mapping. The SCELA results from a demand for self-consistency in the sense of equating the structural mapping and the thermodynamic mapping.

d. Generalized Effective Liquid Theory. A further development of this idea was proposed by Lutsko and Baus [25, 39]. It extends the idea of the SCELA by requiring self-consistency for all densities along the integration path in the hope that this would suppress the contribution of higher order terms. The result is

$$\frac{1}{V}\beta F_{ex}[\rho_1] = -\frac{1}{V}\int_0^1 d\lambda \int d\mathbf{r}_1 d\mathbf{r}_2 (1-\lambda) c_2(r_{12}; \bar{\rho}_{GELA}(\lambda))\rho_1(\mathbf{r}_1)\rho_1(\mathbf{r}_2) + \dots$$

with

$$(76)$$

$$\begin{aligned}
\beta\psi_{ex}(\bar{\rho}_{GELA}(\alpha)) &= \frac{1}{\alpha\rho_1 V}\beta F_{ex}[\alpha\rho_1] \\
&= -\frac{1}{\alpha\bar{\rho}_1 V}\int_0^1 d\lambda \int_0^\lambda d\lambda' \int d\mathbf{r}_1 d\mathbf{r}_2 c_2(\mathbf{r}_{12}; \bar{\rho}_{GELA}(\alpha\lambda')) \\
&\quad \times \alpha\rho_1(\mathbf{r}_1)\alpha\rho_1(\mathbf{r}_2) + \dots \\
&= -\frac{1}{\alpha\bar{\rho}_1 V}\int_0^\alpha d\gamma \int_0^\alpha d\lambda' \int d\mathbf{r}_1 d\mathbf{r}_2 c_2(\mathbf{r}_{12}; \bar{\rho}_{GELA}(\gamma')) \\
&\quad \times \rho_1(\mathbf{r}_1)\rho_1(\mathbf{r}_2) + \dots
\end{aligned}$$

$$(77)$$

Notice that when truncated at second order, this can be converted into a differential equation for $\bar{\rho}_{GELA}(\alpha)$ [25],

$$\frac{\partial^2}{\partial\alpha^2}\alpha\beta\psi_{ex}(\bar{\rho}_{GELA}(\alpha)) = -\frac{1}{\bar{\rho}_1 V}\int d\mathbf{r}_1 d\mathbf{r}_2 c_2(r_{12}; \bar{\rho}_{GELA}(\alpha))\rho_1(\mathbf{r}_1)\rho_1(\mathbf{r}_2) \quad (78)$$

e. Modified Weighted Density Approximation. The Weighted Density Approximation (WDA) of Curtin and Ashcroft [40] will be described separately below. The goal was to try to construct a density functional having the property that it reproduces the DCF of the liquid in the uniform limit,

$$c_2(r_{12}; \bar{\rho}) = -\lim_{\rho(\mathbf{r})\to\bar{\rho}} \frac{\delta^2\beta F^{(WDA)}[\rho]}{\delta\rho(\mathbf{r}_1)\delta\rho(\mathbf{r}_2)} \quad (79)$$

The Modified Weighted Density Approximation of Denton and Ashcroft is a simplified form of the same idea [41]. It is derived by introducing an effective liquid ansatz,

$$\frac{1}{\bar{\rho} V}F_{ex}[\rho] = \psi_{ex}(\bar{\rho}_{MWDA}[\rho]) \quad (80)$$

and writing the MWDA density as a weighted average,

$$\bar{\rho}_{MWDA}[\rho] = \frac{1}{\bar{\rho} V}\int w(r_{12}; \bar{\rho}_{MWDA}[\rho])\rho(\mathbf{r}_1)\rho(\mathbf{r}_2) d\mathbf{r}_1 d\mathbf{r}_2 \quad (81)$$

and demanding that the weighting function be normalized. Demanding that the theory reproduce a given DCF, $c_2(r_{12}; \bar{\rho})$, in the bulk limit is enough to uniquely determine the weight function as

$$w(r_{12}; \bar{\rho}) = \frac{-1}{2\beta\psi'(\bar{\rho})} \left(c_2(r_{12}; \bar{\rho}) + \frac{1}{V}\bar{\rho}\beta\psi''(\bar{\rho}) \right) \tag{82}$$

Then the effective density is determined from

$$2\bar{\rho}_{MWDA}\beta\psi'(\bar{\rho}_{MWDA}) + \bar{\rho}_1\bar{\rho}_{MWDA}\beta\psi''(\bar{\rho}_{MWDA})$$
$$= -\frac{1}{\bar{\rho}V}\int c_2(r_{12}; \bar{\rho}_{MWDA})\rho(\mathbf{r}_1)\rho(\mathbf{r}_2)d\mathbf{r}_1 d\mathbf{r}_2 \tag{83}$$

This is equivalent to a perturbative theory, Eqs. (44) and (45), truncated at second order with $\bar{\rho}(\lambda) = \bar{\rho}_{MWDA}$ and $\bar{\rho}_0$ determined from

$$\bar{\rho}\psi_{ex}(\bar{\rho}_0) + (\bar{\rho}-\bar{\rho}_0)\bar{\rho}_0\psi'_{ex}(\bar{\rho}_0) = \bar{\rho}\psi_{ex}(\bar{\rho}_{MWDA})$$
$$-(\bar{\rho}\bar{\rho}_{MWDA}-2\bar{\rho}_0\bar{\rho}+\bar{\rho}_0^2)\psi'_{ex}(\bar{\rho}_{MWDA})$$
$$-\frac{1}{2}(\bar{\rho}-\bar{\rho}_0)^2\bar{\rho}_{MWDA}\psi''_{ex}(\bar{\rho}_{MWDA}) \tag{84}$$

2. Nonlocal Theories

a. Simple Position-Dependent Effective Liquid. In the simple truncated perturbation theory, one effectively replaces the DCF of the inhomogeneous system by that of a fluid at some specified density. For bulk properties, this might suffice, even for the bulk solid, but it obviously runs into conceptual difficulties when applied to more complex systems. For example, what single value of the reference density should be chosen to approximate correlations in a liquid–vapor interface or vapor–solid interface? For a liquid–vapor system, far from the interface one knows what the DCF should be (since the DCF in the bulk is required as input for most of the DFTs discussed here). This suggests the use of a position-dependent reference density. Since the DCF is a two-point function that is symmetric in its arguments, some care must be taken in introducing such an approximation. Two simple possibilities are

$$c_2(\mathbf{r}_1, \mathbf{r}_2; [\rho]) \simeq c_2\left(r_{12}; \frac{\rho(\mathbf{r}_1)+\rho(\mathbf{r}_2)}{2} \right) \tag{85}$$

and

$$c_2(\mathbf{r}_1, \mathbf{r}_2; [\rho]) \simeq \frac{1}{2}(c_2(r_{12}; \rho(\mathbf{r}_1)) + c_2(r_{12}; \rho(\mathbf{r}_2))) \tag{86}$$

Substituting the first into the exact expression, Eq. (44), gives, after some rearrangement,

$$
\beta F_{ex}[\rho] \simeq \int d\mathbf{r}\, \beta f_{ex}(\rho(\mathbf{r}))
$$
$$
+ \frac{1}{4}\int_v d\mathbf{r}_1 d\mathbf{r}_2 (\rho(\mathbf{r}_1)-\rho(\mathbf{r}_2))^2 \bar{\bar{c}}_2\left(r_{12}; \frac{\rho(\mathbf{r}_1)+\rho(\mathbf{r}_2)}{2}, \bar{\rho}_0\right)
$$
$$
- \frac{1}{2}\int_v d\mathbf{r}_1 d\mathbf{r}_2 \left[\left(\frac{\rho(\mathbf{r}_1)+\rho(\mathbf{r}_2)}{2}-\bar{\rho}_0\right)^2 \bar{c}_2\left(r_{12};\frac{\rho(\mathbf{r}_1)+\rho(\mathbf{r}_2)}{2}, \bar{\rho}_0\right) \right.
$$
$$
\left. -(\rho(\mathbf{r}_1)-\bar{\rho}_0)^2 \bar{\bar{c}}_2(r_{12};\rho(\mathbf{r}_1),\bar{\rho}_0) \right]
$$
$$(87)$$

for the first approximation, where

$$
\bar{\bar{c}}_2(r;\rho,\bar{\rho}_0) \equiv 2\int_0^1 \int_0^\lambda c_2(r;\bar{\rho}_0+\lambda'(\rho-\bar{\rho}_0))d\lambda'd\lambda \qquad (88)
$$

The second approximation gives a somewhat simpler result,

$$
\beta F_{ex}[\rho] = \int d\mathbf{r}\, f_{ex}(\rho(\mathbf{r}))
$$
$$
+ \frac{1}{2}\int_v (\rho(\mathbf{r}_1)-\bar{\rho}_0)(\rho(\mathbf{r}_1)-\rho(\mathbf{r}_2))\bar{c}_2(r_{12};\rho(\mathbf{r}_1),\bar{\rho}_0)d\mathbf{r}_2 d\mathbf{r}_1 \qquad (89)
$$

which has the intuitively appealing form of a local effective liquid approximation plus a contribution that depends on density gradients. It is straightforward to show that an expansion in the inhomogeneity—that is, in powers of $(\rho(\mathbf{r}_1)-\rho(\mathbf{r}_2))$—gives [42]

$$
\beta F_{ex}[\rho] = \int d\mathbf{r}\, f(\rho(\mathbf{r})) + \frac{1}{4}\int_v (\rho(\mathbf{r}_1)-\rho(\mathbf{r}_2))^2
$$
$$
\times \left[2\int_0^1 \lambda c_2\left(r_{12};\bar{\rho}_0+\lambda\left(\frac{\rho(\mathbf{r}_1)+\rho(\mathbf{r}_2)}{2}-\bar{\rho}_0\right)\right)d\lambda \right] d\mathbf{r}_1 d\mathbf{r}_2 + \cdots
$$
$$(90)$$

and further expanding the integrand about $\lambda = 1$ gives, to lowest order,

$$
\beta F_{ex}[\rho] = \int d\mathbf{r}\, f(\rho(\mathbf{r}))
$$
$$
+ \frac{1}{4}\int_V (\rho(\mathbf{r}_1)-\rho(\mathbf{r}_2))^2 c_2\left(r_{12};\frac{\rho(\mathbf{r}_1)+\rho(\mathbf{r}_2)}{2}\right)d\mathbf{r}_1 d\mathbf{r}_2 + \cdots
$$
$$(91)$$

which is the well-known form used by Saam and Ebner in some of the earliest DFT calculations [7, 8].

b. *Weighted Density Approximation of Curtin and Ashcroft.* One criticism of many of the models discussed so far is that they are formally inconsistent. They require as input the DCF of the bulk liquid, but the second functional derivative of the model excess free energy functional does not, in the uniform limit, give the input function $c_2(r_{12}; \rho)$. The Weighted Density Approximation of Curtin and Ashcroft [40] was specifically designed to solve this problem. They begin by writing the exact expression for the excess free energy functional, Eq. (44), with $\bar{\rho}_0 = 0$ as

$$F_{ex}[\rho] = \int \Psi_{ex}(\mathbf{r}; [\rho]) \rho(\mathbf{r}) d\mathbf{r} \tag{92}$$

where

$$\Psi_{ex}(\mathbf{r}; [\rho]) = -\int_0^1 d\lambda \int_0^\lambda d\lambda' \int d\mathbf{r}_2 c_2(\mathbf{r}_1, \mathbf{r}_2; [\lambda'\rho_1]) \rho_1(\mathbf{r}_2)$$

They then introduce a *local* effective liquid approximation

$$\Psi_{ex}(\mathbf{r}; [\rho]) = \psi_{ex}(\rho_{WDA}(\mathbf{r}; [\rho])) \tag{93}$$

where we recall that $\psi(\bar{\rho}) = \frac{1}{\bar{\rho}V} F(\bar{\rho})$ is the free energy per atom and $\psi_{ex}(\bar{\rho})$ is the excess contribution. The local effective density is expressed in terms of a weighted-density ansatz,

$$\rho_{WDA}(\mathbf{r}_1; [\rho]) = \int w(\mathbf{r}_1 - \mathbf{r}_2; \rho_{WDA}(\mathbf{r}_1; [\rho])) \rho(\mathbf{r}_2) d\mathbf{r}_2 \tag{94}$$

The weighting function is fixed by demanding that it is normalized, $\int w(\mathbf{r}; \bar{\rho}) d\mathbf{r} = 1$, and that the ansatz be consistent with the (known) DCF in the liquid state,

$$c_2(r_{12}; \bar{\rho}) = -\lim_{\rho(\mathbf{r}) \to \bar{\rho}} \frac{\delta^2 \beta F_{ex}[\rho]}{\delta\rho(\mathbf{r}_1)\delta\rho(\mathbf{r}_2)} \tag{95}$$

leading to [40] $w(\mathbf{r}; \bar{\rho}) = w(r; \bar{\rho})$ and an integrodifferential equation for $w(r; \bar{\rho})$. Taking the Fourier transform,

$$\tilde{w}(k; \bar{\rho}) = \int \exp(i\mathbf{k} \cdot \mathbf{r}) w(r; \bar{\rho}) d\mathbf{r} \tag{96}$$

the weighting function is determined from

$$k_B T \tilde{c}_2(k; \bar{\rho}) = -2 \frac{\partial \psi_{ex}(\bar{\rho})}{\partial \bar{\rho}} \tilde{w}(k; \bar{\rho}) - \bar{\rho} \frac{\partial^2 \psi_{ex}(\bar{\rho})}{\partial \bar{\rho}^2} \tilde{w}^2(k; \bar{\rho})$$
$$- 2\bar{\rho} \frac{\partial \psi_{ex}(\bar{\rho})}{\partial \rho} \tilde{w}(k; \bar{\rho}) \frac{\partial \tilde{w}(k; \bar{\rho})}{\partial \bar{\rho}} \tag{97}$$

The WDA is therefore computationally more complex than the simpler effective liquid theories because it involves a different effective liquid for each wave vector.

c. Tarazona's Weighted Density Theory. A similar theory was constructed by Tarazona [43] for the specific case of hard spheres. However, rather than enforce the exact relation between the excess free energy functional and the DCF, a simpler approximation was employed. The structure of the theory is the same as that of the WDA, but the determination of the weighting function is different. It is expanded as a series,

$$w(r; \rho) = w_0(r) + w_1(r)\rho + w_2(r)\rho^2 \tag{98}$$

and the first two functions, $w_0(r)$ and $w_1(r)$, are determined by requiring agreement with the first two terms of the virial expansion of the DCF. The final function is fit so as to give a reasonable reproduction of the Percus–Yevick DCF at higher densities.

C. Some Applications

1. Freezing of Hard Spheres

The equilibrium density distribution under action of a given external potential ϕ_0, fixed chemical potential μ and inverse temperature β is that which minimizes the free energy functional $\Omega[\rho; \phi_0]$. In the language of DFT, different phases correspond to different density distributions. Two phases with densities $\rho_1(\mathbf{r})$ and $\rho_2(\mathbf{r})$ can coexist when they simultaneously minimize this functional—that is, when they both satisfy the Euler–Lagrange equation and when $\Omega[\rho_1; \phi_0] = \Omega[\rho_2; \phi_0]$. Since they satisfy they minimize the free energy functional, its value at those density distributions is the grand-canonical free energy, $\Omega[\rho_1; \phi_0] = \Omega$. Since the grand potential is proportional to the pressure, $\Omega = -PV$, this implies the usual thermodynamic condition of equal pressures. The Euler–Lagrange equation,

$$\frac{\delta F[\rho_1]}{\delta \rho_1(\mathbf{r})} = \mu - \phi_0(\mathbf{r}) = \frac{\delta F[\rho_2]}{\delta \rho_2(\mathbf{r})} \tag{99}$$

generalizes the usual condition of equal chemical potentials. Thus, in the grand ensemble used in DFT, two-phase coexistence is in principle determined by adjusting the parameter μ, determining the resultant densities $\rho_1(\mathbf{r})$ and $\rho_2(\mathbf{r})$ and evaluating $\Omega[\rho_1; \phi_0]$ and $\Omega[\rho_2; \phi_0]$ until a value of μ is found for which $\Omega[\rho_1; \phi_0] = \Omega[\rho_1; \phi_0]$. For uniform phases, say liquid and vapor, with a field that only acts to confine the system to a large but finite volume, this reduces to the usual conditions:

$$\frac{\partial f(\bar{\rho}_1)}{\partial \bar{\rho}_1} = \mu = \frac{\partial f(\bar{\rho}_2)}{\partial \bar{\rho}_2}$$

$$f(\bar{\rho}_1) - \mu = f(\bar{\rho}_2) - \mu \tag{100}$$

Solids are most often modeled by using the Gaussian parameterization of the density [see Eqs. (54) and (55) and the accompanying discussion]. In this case, the parameters are the number density of lattice sites, $\bar{\rho}_{\text{latt}}$ (or, equivalently, the lattice constant a), the Gaussian parameter α, and the occupancy x; or, equivalently, the average density is $\bar{\rho} = x\bar{\rho}_{\text{latt}}$. Assuming no field except for at the boundaries of the (large) volume, the Euler–Lagrange equations for the uniform solid are

$$\frac{\partial F[\rho]}{\partial \bar{\rho}_{\text{latt}}} = \mu x$$

$$\frac{\partial F[\rho]}{\partial x} = \mu \bar{\rho}_{\text{latt}} \tag{101}$$

$$\frac{\partial F[\rho]}{\partial \alpha} = 0$$

In fact, in many calculations, the occupancy is held fixed to $x = 1$ since one expects values very close to this in equilibrium solids. In this case, the second of these equations is dropped.

All of the theories discussed above give reasonable results for hard-sphere freezing. Obviously, the quality of the numerical predications depends on the quality of the liquid-state data that all of these theories require (i.e., the DCF of the liquid). The Percus–Yevick DCF is very good at low densities but inaccurate at the high densities characteristic of freezing. For some theories such as the MWDA, SCELA, and GELA, this is not important for the solid phase as the effective densities that enter these theories tend to be about half the actual density. However, in all cases the liquid thermodynamics must also be evaluated, and this always requires densities at which the Percus–Yevick approximation is not very good. Since all one really needs for the liquid is the equation of state, it is common in these cases to use the Percus–Yevick approximation for the solid and to use the Carnahan–Starling approximation for the liquid because both can be understood as being accurate approximations in the relevant domains. For other DFTs that require the liquid-state DCF at all densities, such as RY, ELA, and WDA, this would be inconsistent and one must either suffice with Percus–Yevick or make use of the more accurate, but more complex, parameterization of Hendersen and Grundke [44] or the semi-phenomenological approximation of Baus and Colot [45].

The results of several calculations for liquid-FCC solid hard-sphere coexistence are shown in Table I. Rather than the value of the Gaussian parameter, the more physical Lindemann parameter L, defined as mean-squared displacement divided by the lattice constant, is reported. For the Gaussian model, one has $L = (3/\alpha a^2)^{1/2}$ for the FCC solid and $L = (2/\alpha a^2)^{1/2}$ for the BCC solid. All of the theories give reasonable results, with the GELA being the closest to the

TABLE I

Comparison of the Predictions of Various Effective-Liquid DFTs for the Freezing of Hard Spheres to Data from Simulation[a]

Theory	EOS	$\bar{\eta}_{\text{liq}}$	$\bar{\eta}_{\text{sol}}$	P^*	L
RY[b]	PY	0.506	0.601	15.1	0.06
MWDA[c]	CS	0.476	0.542	10.1	0.097
ELA[d]	PY	0.520	0.567	16.1	0.074
SCELA[e]	CS	0.508	0.560	13.3	0.084
GELA[e]	CS	0.495	0.545	11.9	0.100
WDA[c,f]	CS	0.480	0.547	10.4	0.093
MC[g]	—		0.494	0.545	11.7	0.126

[a]Given are the liquid ($\bar{\eta}_{\text{liq}}$) and solid ($\bar{\eta}_{\text{sol}}$) packing fractions ($\eta = \pi \rho d^3 / 6$), the reduced pressure ($P^* = \beta P d^3$), and the Lindemann parameter (L) at bulk coexistence. For each theory, the equation of state used for the fluid, Percus–Yevick (PY), or Carnahan–Starling (CS) is indicated.
[b]From Barrat et al. [49].
[c]From Denton and Ashcroft [41].
[d]From Baus and Colot [37].
[e]From Lutsko and Baus [25].
[f]From Curtin and Ashcroft [40].
[g]From Hoover and Ree [57].

simulation values. One important defect shared by all theories is that they uniformly predict a value of the Lindemann parameter which is much too low. The significance of the apparent accuracy in describing some other properties at coexistence is therefore open to question. It is also possible to study freezing into other structures, such as the BCC and simple cubic (SC) lattices [25]. However, since these structures are metastable, little information is available with the exception of Curtin and Runge, who used a constrained simulation method to study hard spheres in a BCC configuration [46]. At $\bar{\rho} d^3 = 1.041$ they found an excess free energy relative to a uniform ideal gas of $\frac{1}{\bar{\rho} V} \beta F(\rho) - (\ln \bar{\rho} \Lambda^3 - 1) = 6.094$ and at $\bar{\rho} d^3 = 1.1$ they found the value 6.878. The WDA gives the values 5.975 and 6.771, respectively, while the GELA gives 6.118 and 6.991 using the PY DCF and 6.049 and 6.903 using the CS DCF [25]. This seems to confirm the trends seen in the FCC coexistence data. The accuracy of the various theories in predicting the pressure of the solid phase for all densities follows the same trends, with the GELA being very close to simulation. At high densities, problems develop with the MWDA where multiple solutions to the effective-density equation develop and where there are regions of no solution [47, 48]. At very high densities, the GELA seems to predict that the Lindemann parameter goes to zero at $\eta \simeq 0.736$, which is very near the FCC close packing density, $\eta = 0.74$, where the Lindemann parameter must be zero. However, for the BCC phase, the GELA predicts a Lindemann parameter that not only varies little with density, showing no sign of going to zero at BCC close packing, $\eta = 0.68$, but also even increases somewhat with increasing density. All of this is highly unphysical and indicates a breakdown of the theory.

This generally satisfactory behavior does not translate to other systems. Barrat, Hansen, Pastore, and Waisman [49] used the RY and ELA theories to investigate the freezing of atoms interacting with an inverse-power potential, $v(r) = (\sigma/r)^n$. In the limit $n \to \infty$, these so-called "soft spheres" coincide with the usual hard spheres. For the DCF of the fluid, they used both the Rogers and Young integral equation [50] and the modified HNC scheme of Rosenfeld and Ashcroft [51]. Their conclusions were that neither theory stabilized the BCC phase even though from simulation, it is known that the BCC phase is the stable solid phase for $n \le 6$ [52]. Laird and Kroll carried out a similar study using the MWDA, SCELA and GELA [53]. They found that the MWDA also predicted freezing into the FCC structure for all values of n with the numerical values of the freezing parameters worsening with decreasing n. Worse yet, the SCELA and GELA failed to predict freezing altogether for $n \le 6$. De Kuijper, Vos, Barrat, Hansen, and Schouten investigated the ability of these theories to predict freezing for many other potentials including the Lennard-Jones and exponential-6 potentials often used to model simple liquids [54]. They found that the SCELA fails to predict freezing for the LJ potential while the MWDA fails at low temperatures but does predict freezing into an FCC structure at higher temperatures. The simple second-order perturbation theory predicts freezing but with poor values for the freezing parameters. The conclusion is that none of these theories appears to be reliably predictive as one moves away from the hard-sphere potential. Further details including a discussion of freezing in binary systems can be found in Ref. (55). There have been attempts to further modify these basic theories to give a better description of freezing. For example, Wang and Gast suggested modifying the MWDA prescription for the effective density so as to give the correct static-lattice free energy in the large α limit [56]. While this and other attempts have yielded significant improvements, the trend in recent years has been toward simpler theories that separate the hard core and attractive contributions to the free energy as discussed below.

2. Liquid–Solid Interface

These theories have also been used to study interfacial systems including fluids near walls and liquid–solid interfaces. We note in particular the work by Curtin [58], Ohensorge, Löwen, and Wagner [59, 60] and Marr and Gast [61] on the liquid–solid interface in Lennard-Jones systems, as well as the work of Kyrlidis and Brown [62] on the hard-sphere liquid–solid interface, all of which demonstrate the utility of DFT even in describing very inhomogeneous systems. A good summary of the work on wetting can be found in Evans [14]. Numerous applications of interest in chemical engineering can be found in Wu [17].

3. Properties of the Bulk Liquid State

One interesting application of DFT is to model correlations in the liquid state. The third- and higher-order DCFs can be calculated from any model DFT by functional differentiation. For example, with the MWDA theory,

$$F_{ex}[\rho] = N\psi_{ex}(\bar{\rho}_{MWDA}[\rho])$$

$$\bar{\rho}_{MWDA} = \frac{1}{N}\int w(r_{23}; \bar{\rho}_{MWDA})\rho(\mathbf{r}_2)\rho(\mathbf{r}_3)d\mathbf{r}_2 d\mathbf{r}_3 \qquad (102)$$

one has that

$$c_1(r; \bar{\rho}) = -\lim_{\rho(\mathbf{r}) \to \bar{\rho}} \frac{\delta\beta F_{ex}[\rho]}{\delta\rho(\mathbf{r})}$$

$$= -\lim_{\rho(\mathbf{r}) \to \bar{\rho}} \left(\beta\psi_{ex}(\bar{\rho}_{MWDA}[\rho]) + N\frac{\delta\bar{\rho}_{MWDA}[\rho]}{\delta\rho(\mathbf{r})}\psi'_{ex}(\bar{\rho}_{MWDA}[\rho]) \right)$$

$$(103)$$

Implicit differentiation of the equation for the effective density gives

$$N\frac{\delta}{\delta\rho(\mathbf{r}_1)}\bar{\rho}_{MWDA} = \frac{2\int w(r_{12}; \bar{\rho}_{MWDA})\rho(\mathbf{r}_2)d\mathbf{r}_2 - \bar{\rho}_{MWDA}}{1 - \frac{1}{N}\int \frac{\partial}{\partial\bar{\rho}_{MWDA}}w(r_{23}; \bar{\rho}_{MWDA})\rho(\mathbf{r}_2)\rho(\mathbf{r}_3)d\mathbf{r}_2 d\mathbf{r}_3} \qquad (104)$$

so that in the thermodynamic limit, in which the $1/N$ term in the denominator is negligible, the result is

$$c_1(r; \bar{\rho}) = -\lim_{\rho(\mathbf{r}) \to \bar{\rho}} \frac{\delta\beta F_{ex}[\rho]}{\delta\rho(\mathbf{r})}$$

$$= -(\beta\psi_{ex}(\bar{\rho}) + \psi'_{ex}(\bar{\rho})\bar{\rho}) \qquad (105)$$

because the MWDA weight function is normalized, $\int w(r; \bar{\rho})d\mathbf{r} = 1$. This is the usual equilibrium result. In the uniform limit, the second functional derivative of the MWDA (and WDA) functionals give, by construction, the bulk DCF used to calculate the weighting function. The third-order DCF for hard spheres has been evaluated and compared to simulation, with some success using both the WDA [63] and the MWDA [64] models.

There is also a close connection between DFT and the hypernetted chain (HNC) approximation to classical liquid-state theory because, as noted by Kim and Jones [65] and separately by White and Evans [66], the MWDA implies the HNC for a bulk fluid. This is easily seen by considering the DFT calculation of the structure of the bulk fluid. As explained above, the PDF can be obtained by solving for the density profile generated by a particle fixed at the origin which acts as the external field. In principle, the effective density is determined by the coupled equations

$$F_{ex}[\rho] = N\psi_{ex}(\bar{\rho}_{MWDA}[\rho])$$

$$\bar{\rho}_{MWDA} = \frac{1}{N}\int w(r_{23};\bar{\rho}_{MWDA})\rho(\mathbf{r}_2)\rho(\mathbf{r}_3)d\mathbf{r}_2 d\mathbf{r}_3 \tag{106}$$

$$\rho(\mathbf{r};[v]) = \exp\left(\beta\mu - \beta v(r) - \frac{\delta\beta F_{ex}[\rho]}{\delta\rho(\mathbf{r};[v])}\right)$$

where $v(r)$ is the pair potential. However, rewriting the equation for the effective density as

$$\bar{\rho}_{MWDA} = \bar{\rho} + \frac{1}{N}\int w(r_{23};\bar{\rho}_{MWDA})(\rho(\mathbf{r}_2)-\bar{\rho})(\rho(\mathbf{r}_3)-\bar{\rho})d\mathbf{r}_2 d\mathbf{r}_3 \tag{107}$$

it is clear that the difference from the bulk density will be of order $1/N$, provided that the spatial density profile approaches the bulk density sufficiently quickly as one moves away from the center of the external field. Assuming this is the case, then in the thermodynamic limit $\bar{\rho}_{MWDA} = \bar{\rho}$, just as in the case of zero field. To solve for the density profile, the functional derivative of the MWDA excess free energy functional is also needed, and one finds using Eq. (104) that

$$\frac{\delta}{\delta\rho(\mathbf{r})}F_{ex}[\rho] = \frac{\delta}{\delta\rho(\mathbf{r})}N\psi_{ex}(\bar{\rho}_{MWDA}[\rho])$$

$$= \psi_{ex}(\bar{\rho}_{MWDA}[\rho]) + N\frac{\delta\bar{\rho}_{MWDA}[\rho]}{\delta\rho(\mathbf{r})}\psi'_{ex}(\bar{\rho}_{MWDA}[\rho])$$

$$= \psi_{ex}(\bar{\rho}_{MWDA}[\rho]) + \frac{2\psi'_{ex}(\bar{\rho}_{MWDA}[\rho])\int w(r_{12};\bar{\rho}_{MWDA})\rho(\mathbf{r}_2)d\mathbf{r}_2 - \bar{\rho}_{MWDA}}{1 - \frac{1}{N}\int\frac{\partial}{\partial\bar{\rho}_{MWDA}}w(\mathbf{r}_{23};\bar{\rho}_{MWDA})\rho(\mathbf{r}_2)\rho(\mathbf{r}_3)d\mathbf{r}_2 d\mathbf{r}_3}$$

$$\tag{108}$$

In the thermodynamic limit, the term proportional to $1/N$ in the denominator of the second term on the right does not contribute. Using the explicit form of the MWDA weight function, Eq. (82) and a little algebra gives

$$\frac{\delta}{\delta\rho(\mathbf{r})}F_{ex}[\rho] = \psi_{ex}(\bar{\rho}) - \int \beta^{-1}c_2(r_{12};\bar{\rho})\rho(\mathbf{r}_2)d\mathbf{r}_2 - \bar{\rho}^2\psi''(\bar{\rho}) - \bar{\rho}\psi'_{ex}(\bar{\rho})$$

$$= \psi_{ex}(\bar{\rho}) + \bar{\rho}\psi'_{ex}(\bar{\rho}) - \int \beta^{-1}c_2(r_{12};\bar{\rho})(\rho(\mathbf{r}_2)-\bar{\rho})d\mathbf{r}_2 \tag{109}$$

Noting that the bulk density is determined by the Euler–Lagrange equation which reduces to $\ln \bar{\rho} + \psi_{\text{ex}}(\bar{\rho}) + \bar{\rho}\psi'_{\text{ex}}(\bar{\rho}) = \mu$, the final result is

$$\rho(\mathbf{r}; [v]) = \bar{\rho}\exp\left(\beta\mu - \beta v(r) + \int c_2(r_{12}; \bar{\rho})(\rho(\mathbf{r}_2; [v]) - \bar{\rho})d\mathbf{r}_2\right) \quad (110)$$

Thus, for the uniform system, one has

$$g(r_1; \mu) = \exp\left(-\beta v(r_1) + \bar{\rho}\int c_2(r_{12}; \bar{\rho})(g(r_2; \mu) - 1)d\mathbf{r}_2\right) \quad (111)$$

This is the HNC closure relation which, combined with the Ornstein–Zernike equation gives a complete theory of the bulk liquid state [21].

White and Evans also show that the theories are equivalent for a fluid near a wall, but not for other more confined geometries [66]. Denton and Ashcroft reversed the logic and used DFT to give new closures for the Ornstein–Zernike equation [67]. They also note [68] that Barrat, Hansen, and Pastore had previously shown that the RY theory also implies the HNC for a uniform fluid [69].

IV. FUNDAMENTAL MEASURE THEORY FOR HARD SPHERES

Fundamental Measure Theory (FMT) has proven to be one of the most successful methods of modeling the DFT of hard spheres in more than one dimension. It is strongly motivated by Percus' results for the one-dimensional hard-sphere fluid as well as by the Scaled Particle Theory approach to liquid-state theory [10–12].

A. Motivation

1. Hard Rods

In many interesting circumstances, the statistical mechanics of hard spheres in one dimension—usually called hard rods—can be solved exactly [70]. Beginning in the 1970s, Percus cast the problem in the language of Density Functional Theory and gave the solution for arbitrary external field. This result has served as a touchstone in recent developments of DFT for hard spheres. Because it gives an intrinsically interesting example of exact DFT, Percus' solution is outlined in Appendix A while the results are summarized here to provide motivation for the discussion of FMT below.

The great simplification of hard rods is that they only interact with nearest neighbors and they cannot move past one another. Thus, the grand partition function for rods of length d is

$$\Xi(\beta, \mu; [\phi]) = 1 + \int_{-\infty}^{\infty} \exp(-\beta\tilde{\phi}(q_1))dq_1$$
$$+ \sum_{n=2}^{\infty} \int_{-\infty}^{\infty} \exp\left(-\beta\sum_{i=1}^{n}\tilde{\phi}(q_i)\right)W(q_1, \ldots, q_n)dq_1 \ldots dq_n \quad (112)$$

where $\phi(x)$ is the external field, $\tilde{\phi}(x) = \phi(x) - \mu$, and $W(q_1, \ldots, q_n)$ is one provided that $q_1 < q_2 - d < q_3 - 2d \ldots$ and zero otherwise. Functional differentiation with respect to the field gives an expression for the local density, $\rho(x)$. As shown in Appendix A it is possible to eliminate the field in favor of the density thus arriving at

$$\ln\Xi(\beta, \mu; [\rho]) = \int_{-\infty}^{\infty} \frac{\frac{1}{2}(\rho(r+d/2) + \rho(r-d/2))}{1 - \int_{-d/2}^{d/2} \rho(r+y)dy} \, dr \qquad (113)$$

This result is not directly useful for DFT because the density that appears in it is the equilibrium density: The field has been eliminated so that this is the equivalent of Eq. 23. However, as in the simple examples of exact DFT from the previous section, the relation between the field and the density can be used together with the the Euler–Lagrange equation to get

$$\frac{\delta\beta F[\rho]}{\delta\rho(\mathbf{r})} = -\beta\tilde{\phi}(r)$$

$$= \ln\rho(r) - \frac{1}{2}\ln\left(1 - \int_{r-d}^{r}\rho(y)dy\right)dr - \frac{1}{2}\ln\left(1 - \int_{r}^{r+d}\rho(y)dy\right)$$

$$+ \frac{1}{2}\int_{-r-d/2}^{r+d/2}\left(\frac{\rho(x+d/2) + \rho(x-d/2)}{1 - \int_{-d/2}^{d/2}\rho(x+y)dy}\right)dr \qquad (114)$$

Functional integration as described in the last section gives the final result

$$\beta F[\rho] = \beta F_{\text{id}}[\rho] - \int \frac{1}{2}(\rho(x+d/2) + \rho(x-d/2))\ln\left(1 - \int_{-d/2}^{d/2}\rho(x+y)dy\right)dx \qquad (115)$$

As shown below, this beautiful, exact result plays a fundamental role in the construction of FMT.

2. Generalization to Higher Dimensions

In fact, Percus and co-workers speculated on how this might be generalized to more than one dimension [71–73]. It has the form of an integral over a local excess free energy that clearly depends on two quantities: the density evaluated at the "surface" of a hard rod centered at position x, $\rho(x \pm d/2)$, and the density averaged over the "volume" of a hard rod centered at position x; these concepts are easily generalized to more than one dimension as will be seen below. This suggests writing it in the form

$$\beta F_{ex}[\rho] = -\int_{-\infty}^{\infty} s(r)\ln(1-\eta(r))dr \tag{116}$$

where

$$\eta(r) = \int_{-\infty}^{\infty} w_\eta(r-y)\rho(y)dy$$

$$s(r) = \int_{-\infty}^{\infty} w_s(r-y)\rho(y)dy \tag{117}$$

and the weight functions are $w_\eta(x) = \Theta\left(\frac{d}{2}-|x|\right)$ and $w_s(x) = \delta\left(\frac{d}{2}-|x|\right)$, where $\Theta(x)$ is the step function equal to 1 for $x > 0$ and zero otherwise. The first of these weights restricts integration to the volume of one of the hard rods, while the second restricts integration to its "surface." In this form, the extrapolation to higher dimensions is obvious. In the context of his work on SPT, Rosenfeld realized that another important property of these "fundamental measures" is that their convolution gives the Mayer function for hard spheres $\Theta(d-|x|)$. To see why this is important, let us denote these, and possibly other, linear density functionals collectively as $n_\alpha(r) = \int_{-\infty}^{\infty} w_\alpha(r-y)\rho(y)dy$ and imagine that the excess free energy can be written, by analogy to Eq. (116), as

$$\beta F_{ex}[\rho] = \int \Phi(\{n_\alpha(\mathbf{r})\})d\mathbf{r} \tag{118}$$

for some algebraic function of the measures, Φ. Then it follows that the DCF should be

$$c_2(\mathbf{r}_1, \mathbf{r}_2; [\rho]) = -\frac{\delta^2 \beta F_{ex}[\rho]}{\delta\rho(\mathbf{r}_1)\delta\rho(\mathbf{r}_2)} = -\sum_{\alpha\gamma} \int \frac{\partial^2 \Phi}{\partial n_\alpha(\mathbf{r})\partial n_\gamma(\mathbf{r})} w_\alpha(\mathbf{r}-\mathbf{r}_1)w_\gamma(\mathbf{r}-\mathbf{r}_2)d\mathbf{r} \tag{119}$$

and, at zero density, this must become the negative of the Mayer function. Thus, if the free energy can be written in this form, the convolution of the weights must give a step function.

B. Rosenfeld's FMT

Rosenfeld begins by noting that in three dimensions, the step function can be written as a convolution using three basic functions: $w_i^{(3)}(\mathbf{r}) = \Theta\left(\frac{d_i}{2}-r\right)$, $w_i^{(2)}(\mathbf{r}) = \delta\left(\frac{d_i}{2}-r\right)$, and $\mathbf{w}_i^{(2)}(\mathbf{r}) = \hat{\mathbf{r}}\delta\left(\frac{d_i}{2}-r\right)$, where $\hat{\mathbf{r}} = \mathbf{r}/r$ is the unit vector and the index i is to distinguish different species (i.e., particles with different hard-sphere diameters). Note that the spatial integral of $w^{(3)}$ is the volume of a sphere with diameter d, that of $w^{(2)}$ is its area. This suggests also defining $w_i^{(1)}(\mathbf{r}) = w_i^{(2)}(\mathbf{r})/4\pi(d/2)$ and $w_i^{(0)}(\mathbf{r}) = w^{(2)}(\mathbf{r})/4\pi(d/2)^2$, which integrate to give the radius of the sphere and unity, respectively. They therefore constitute the set of "fundamental measures" of the sphere and are the origin of the name of the theory. Denoting a convolution between two functions $f(\mathbf{r})$ and $g(\mathbf{r})$ as

$$f \otimes g \equiv \int f(\mathbf{r}_1 - \mathbf{r}) g(\mathbf{r} - \mathbf{r}_2) d\mathbf{r} \tag{120}$$

it is easy to confirm that

$$\Theta\left(\frac{d_i + d_j}{2} - r_{12}\right) = w_3^i \otimes w_0^j + w_0^i \otimes w_3^j + w_2^i \otimes w_1^j + w_1^i \otimes w_2^j - \mathbf{w}_2^i \otimes \mathbf{w}_1^j - \mathbf{w}_1^i \otimes \mathbf{w}_2^j \tag{121}$$

where the scalar product is also taken in the final two terms on the right and where $\mathbf{w}_i^{(1)}(\mathbf{r}) = w_i^{(2)}(\mathbf{r})/4\pi(d/2)$. The reason for introducing the somewhat redundant $w_i^{(0)}$ and $w_i^{(0)}$ is that in the uniform limit, the resulting density variables correspond to variables occuring in SPT. If the density distribution of species i is $\rho_i(\mathbf{r})$ then the fundamental densities are defined as

$$n_\alpha(\mathbf{r}) = \sum_i \int w_i^{(\alpha)}(\mathbf{r} - \mathbf{r}_1) \rho(\mathbf{r}_1) d\mathbf{r}_1 \tag{122}$$

Rosenfeld next makes the ansatz that the functional $F[n]$ has the form given in Eq. (118) above. Then, if the density satisfies the Euler–Lagrange equation for some field, then from Eq. (123) we obtain

$$\Omega = F_{\text{ex}}[\rho] - \int \left(k_B T \rho(\mathbf{r}) + \frac{\delta F_{\text{ex}}[\rho]}{\delta \rho(\mathbf{r})}\right) d\mathbf{r}$$

$$= -k_B T \int \left(\rho(\mathbf{r}) - \Phi(\{n_\alpha(\mathbf{r})\}) + \sum_\alpha \frac{\partial \Phi}{\partial n_\alpha(\mathbf{r})} n_\alpha(\mathbf{r})\right) d\mathbf{r} \tag{123}$$

Since the grand potential is just $\Omega = -PV$, in a uniform system with $\rho(\mathbf{r}) = \bar{\rho}$, one has that

$$\beta P = \bar{\rho} - \Phi(\{n_\alpha\}) + \sum_{\alpha,i} \frac{\partial \Phi}{\partial n_\alpha} n_\alpha \tag{124}$$

Another relation is obtained using the ideas of SPT. As explained in Roth et al. [74], the chemical potential for inserting a single spherical particle is equal to the work of insertion. This consists of work done against the fluid pressure, PV, work done due to surface tension, proportional to the area of the sphere, and subdominant terms due to the dependence of surface tension on the radius of curvature. Thus, in the limit that the sphere becomes infinitely large, the work done divided by the volume of the sphere is equal to the pressure, P. Thus, since in the bulk we have

$$\beta\mu_{ex}^{i} = \frac{\partial\Phi}{\partial\rho^{i}} = \frac{\partial\Phi}{\partial n_0} + \frac{\partial\Phi}{\partial n_1}R_i + \frac{\partial\Phi}{\partial n_2}S_i + \frac{\partial\Phi}{\partial n_3}V_i \qquad (125)$$

this reasoning implies that

$$\frac{\partial\Phi}{\partial n_3} = \beta P \qquad (126)$$

Combining this and Eq. (124) gives

$$\frac{\partial\Phi}{\partial n_3} = n_0 - \Phi(\{n_\alpha\}) + \sum_\alpha \frac{\partial\Phi}{\partial n_\alpha}n_\alpha \qquad (127)$$

The final assumption made is that the dependence of Φ on the fundamental densities can be determined by dimensional analysis. Specifically, it is assumed that since Φ has units of inverse volume, it must be of the form

$$\Phi = f_0(n_3)n_0 + f_1(n_3)n_1 n_2 + f_2(n_3)n_2^3 + f_3(n_3)\mathbf{n}_1 \cdot \mathbf{n}_2 + f_4(n_3)n_2(\mathbf{n}_2 \cdot \mathbf{n}_2) \quad (128)$$

In a single-component system, combinations such as $n_1 n_2/n_0$ could occur; however, in a multicomponent system, such terms would not allow a virial expansion in all of the partial densities and so are ruled out. Substituting into Eq. (127) and treating the densities as independent variables gives

$$\begin{aligned}
f_0'(n_3) &= 1 + n_3 f_0'(n_3) \\
f_1'(n_3) &= f_1(n_3) + n_3 f_1'(n_3) \\
f_2'(n_3) &= 2f_2(n_3) + n_3 f_2'(n_3) \\
f_3'(n_3) &= f_3(n_3) + n_3 f_3'(n_3) \\
f_4'(n_3) &= 2f_4(n_3) + n_3 f_4'(n_3)
\end{aligned} \qquad (129)$$

Solution of these equations with the resulting integration constants chosen to give the correct low-density behavior results in the density functional model of Rosenfeld [75, 76]:

$$\Phi = -n_0 \ln(1-n_3) + \frac{n_1 n_2 - \mathbf{n}_1 \cdot \mathbf{n}_2}{1-n_3} + \frac{1}{24\pi}\frac{n_2^3 - 3n_2(\mathbf{n}_2 \cdot \mathbf{n}_2)}{(1-n_3)^2} \qquad (130)$$

(Note that in fact, the vector density measures are zero in a uniform system so that the last two terms in Eq. (128) should not be included [74]. Therefore the result depends on assuming that Eq. (127) continues to hold in a slightly *inhomogeneous* system.)

For pure (single-species) systems, an alternative notation that is commonly used in the literature emphasizes the basis functions rather than the fundamental measures. One defines $w^{(\eta)}(\mathbf{r}) = \Theta(\frac{d}{2}-r)$, $w^{(s)}(\mathbf{r}) = \delta(\frac{d}{2}-r)$ and $\mathbf{w}^{(v)}(\mathbf{r}) = \hat{\mathbf{r}}\delta(\frac{d}{2}-r)$ and denotes the corresponding density measures as $\eta(\mathbf{r}), s(\mathbf{r})$ and $v(\mathbf{r})$, and the free energy density functional is written as

$$\Phi = \Phi_1 + \Phi_2 + \Phi_3 \tag{131}$$

with

$$\Phi_1 = -\frac{1}{\pi d^2} s \ln(1-\eta)$$

$$\Phi_2 = \frac{1}{2\pi d} \frac{s^2 - v^2}{(1-\eta)} \tag{132}$$

$$\Phi_3 = \frac{1}{24\pi} \frac{s^3 - 3sv^2}{(1-\eta)^2}$$

Rosenfeld's model possesses several remarkable properties. The leading term of the free energy density is proportional to the exact functional for one-dimensional hard spheres, which is at least intuitively satisfying. In the bulk liquid, the local density is a constant, $\bar{\rho}$, $\eta = \pi\bar{\rho}d^3/6$ is the usual packing fraction, $s = \pi\bar{\rho}d^2 = 6\eta/d$, and the vector density measures vanish. The resulting expression for the free energy density, which is then just the bulk free energy per unit volume, becomes

$$f_{\text{ex}}(\rho) = \Phi = \bar{\rho}\left(-\ln(1-\eta) + \frac{3}{2}\eta\frac{2-\eta}{(1-\eta)^2}\right) \tag{133}$$

which is recognized as the Percus–Yevick compressibility equation of state [21]; it is of course the same as Eq. (130) without the vector terms, which is the usual result of the simplest form of SPT [10–12]. Finally, evaluating the DCF from Eq. (119) gives

$$c(r;\bar{\rho}) = -\Theta(d-r)\left(\frac{(1+2\eta)^2}{(\eta-1)^4} - 6\eta\frac{(1+\frac{\eta}{2})^2}{(\eta-1)^4}\left(\frac{r}{d}\right) + \frac{\eta}{2}\frac{(1+2\eta)^2}{(\eta-1)^4}\left(\frac{r}{d}\right)^3\right) \tag{134}$$

which again agrees with the Percus–Yevick result. Thus, the Rosenfeld FMT has the property that, at least in the case of hard spheres, it is completely consistent with the Percus–Yevick hard-sphere theory. In many ways, this was, and remains, one of its most attractive features. In fact, an alternative derivation of this functional was given by Kierlik and Rosinberg [77] who, rather than introduce the fundamental measure weights a priori, instead demanded that the SPT equation of state hold (that is, Eq. (130) without the vector terms) and asked what weights one would need in Eq. (122) to get the Percus–Yevick DCF. The resulting theory was subsequently shown to be mathematically identical to Rosenfeld's [78].

C. Refinement of FMT

Despite the early successes, it was also recognized that the Rosenfeld FMT did not give an adequate description of hard-sphere statistical mechanics. The two most important issues had to do with the seemingly independent problems of the solid phase [75] and of the description of reduced dimension systems such as quasi-one-dimensional pores [79].

The calculation of the free energy of the solid phase within the Gaussian approximation requires knowledge of the local densities. From the explicit forms of the weights, it is easy to see that the local densities are related by

$$s(\mathbf{r}; d) = 2 \frac{\partial}{\partial d} \eta(\mathbf{r}; d)$$

$$\mathbf{v}(\mathbf{r}; d) = \nabla \eta(\mathbf{r}; d)$$

(135)

so that once the local packing fraction is calculated, the other densities follow easily. Using the Gaussian model for the densities, straightforward calculation gives

$$\eta(\mathbf{r}) = (\bar{\rho}/\bar{\rho}_{\text{latt}}) \sum_{j=0}^{\infty} \tilde{\eta}(|\mathbf{r} - \mathbf{R}_j|) = \bar{\rho} \sum_{j=0}^{\infty} \exp(-i\mathbf{K}_j \cdot \mathbf{r}) \exp(-K_j^2/2\alpha) \tilde{\tilde{\eta}}(K_j)$$

(136)

where $\bar{\rho}$ is the average density, $\bar{\rho}_{\text{latt}}$ is the density of the lattice sites, the first sum is over lattice vectors, $\mathbf{R}_j$, and the second is over reciprocal lattice vectors, $\mathbf{K}_j$ and

$$\tilde{\eta}(r) = \frac{1}{2} \left(\text{erf}\left(\sqrt{\alpha}\left(r + \frac{d}{2} \right) \right) - \text{erf}\left(\sqrt{\alpha}\left(r - \frac{d}{2} \right) \right) \right)$$

$$- \frac{1}{2r} \sqrt{\frac{1}{\alpha d^2 \pi}} \left(\exp\left(-\alpha\left(r - \frac{d}{2} \right)^2 \right) - \exp\left(-\alpha\left(r + \frac{d}{2} \right)^2 \right) \right)$$

$$\tilde{\tilde{\eta}}(K) = \frac{1}{6} \pi \bar{\rho} d^3 \left(j_0\left(\frac{Kd}{2} \right) + j_2\left(\frac{Kd}{2} \right) \right)$$

(137)

The problem is that solids typically have rather large values of αd^2 (e.g., on the order of 100), and this causes the local packing fraction to be very close to one since (with $\bar{\rho}_{\text{latt}} = \bar{\rho}$)

$$\tilde{\eta}(0) = 1 - \frac{(\alpha d^2)^{1/2}}{\sqrt{\pi}} \exp\left(-\frac{1}{4}\alpha d^2\right) + \cdots \tag{138}$$

leading to possibly large contributions to the free energy since Φ diverges for $\eta = 1$. However, at these points it is also the case that

$$\tilde{s}(0) = \frac{1}{d\sqrt{\pi}} (\alpha d^2)^{3/2} e^{-\frac{1}{4}\alpha d^2} \tag{139}$$

so that the interplay between the various terms must be examined carefully. This was done by Rosenfeld et al. [80], who noted that the Φ_2 contribution to the free energy density did not destabilize the solid because of a cancelation in the numerator between the s^2 and v^2 terms. However, no such cancelation occurs in the Φ_3 term, so the model diverges for highly localized densities. This suggested modifying Φ_3 in such a way that a similar cancelation occurred while at the same time retaining the exact connection to the Percus–Yevick distribution function. They proposed using an expression of the form

$$\Phi_3 = \frac{1}{24\pi} \frac{s^3}{(1-\eta)^2} f(\xi) \tag{140}$$

where $\xi^2 = v^2/s^2$. Setting $f(\xi) = 1-3\xi^2$ corresponds to the original Rosenfeld functional while $f(\xi) = (1-\xi^2)^3$ has all of the desired properties. Another possible choice is $f(\xi) = 1-3\xi^2 + 2\xi^3$. The first modification will be referred to as the RSLT theory, while the second will be called the RSLT2 theory. Both of these modifications stabilize the solid without affecting the properties of the bulk fluid.

D. Dimensional Reduction and the Tarazona Functional

The divergences that occur in the solid are due to the contributions of density distributions that are highly localized at the lattice sites. This eventually led Rosenfeld, Tarazona, and others to study in detail the description of quasi-zero-dimensional systems. The end result was a new approach to the derivation of FMT which points the way to eliminating the divergences that prevent application to the solid phase.

Suppose a field is used which is infinite everywhere except for a cavity, centered at the origin, that is infinitesimally larger than a hard sphere. The average density distribution must therefore be $\rho(\mathbf{r}) = N\delta(\mathbf{r})$, where $0 < N < 1$ is the average occupancy. Consider the first contribution to the Rosenfeld functional written as

$$F_1 = -\int d\mathbf{r}\, \psi_1(\eta(\mathbf{r})) \int ds\, \rho(\mathbf{r}+\mathbf{s}) w_D(s) \qquad (141)$$

where $\psi_1(\eta) = \ln(1-\eta)$, $w_D(s) = S_D^{-1}(d)\,\delta(\frac{d}{2}-s)$, and $S_D(d)$ is the area of the D-sphere with diameter d. Noting that

$$\eta(\mathbf{r}) = \int_{-\infty}^{\infty} \Theta\left(\frac{d}{2} - |\mathbf{r}-\mathbf{r}'|\right)\rho(\mathbf{r}')d\mathbf{r}' = N\Theta\left(\frac{d}{2} - r\right) \qquad (142)$$

and using $\delta(d/2 - x) = 2\frac{\partial}{\partial d}\Theta(d/2 - x)$, one finds

$$F_1 = -S_D^{-1}\int d\mathbf{r}\, \psi_1(\eta(\mathbf{r})) N\delta\left(s - \frac{d}{2}\right) = S_D^{-1}\frac{\partial}{\partial(d/2)}\int \Phi_0(\eta(\mathbf{r}))d\mathbf{r} \qquad (143)$$

where $\Phi_0(\eta)$ is the exact zero-d functional,

$$\Phi_0(\eta) = (1 - \eta)\ln(1 - \eta) - (1 - \eta) \qquad (144)$$

Notice that for $r < d/2$, we have $\eta = N$, while for $r > d/2$ we have $\eta = 0$, and so $\Phi_0(\eta) = 0$. Thus, the integral gives

$$F_1 = S_D^{-1}\frac{\partial}{\partial(d/2)} V_D(d)\Phi_0(N) = \Phi_0(N)$$

which agrees with the exact result, Eq. (37).

Consider next the distribution for a cavity that is the union of two such quasi-zero-dimensional cavities. If the centers are closer than the hard-sphere diameter (so that the two cavities overlap), the combined cavity can still only hold one hard sphere at a time and the exact result is again given by Eq. (37). If one of the cavities is centered at the origin and the other is centered on $\mathbf{R}$, then the density will be $\rho(\mathbf{r}) = N_1\delta(\mathbf{r}) + N_2\delta(\mathbf{r}-\mathbf{R})$ and we again find

$$F_1 = -S_D^{-1}\frac{\partial}{\partial(d/2)}\int \Phi_0(\eta(\mathbf{r}))d\mathbf{r} \qquad (145)$$

Define the region V_1 to be all points such that $r < d/2$ and V_2 to be all points such that $|\mathbf{r}-\mathbf{R}| < d/2$. Let their intersection be $V_{12} = V_1 \cap V_2$ and let $|V_i|$ be the volume of V_i. Then, $\eta(\mathbf{r})$ is N_1 in V_1-V_{12} (i.e., in V_1 but excluding V_{12}) and N_2 in V_2-V_{12} and $N = N_1 + N_2$ in V_{12} so

$$\int \Phi_0(\eta(\mathbf{r}))d\mathbf{r} = |V_1-V_{12}|\Phi_0(N_1) + |V_2-V_{12}|\Phi_0(N_2) + |V_{12}|\Phi_0(N) \qquad (146)$$

Now, specializing to three dimensions, $|V_1| = |V_2| = \frac{4\pi}{3}\left(\frac{d}{2}\right)^3$ while we note for later use that the volume of intersection of two spheres with diameters d_1 and d_2 is

$$|V_{12}| = \frac{\pi}{12}\left(R(d_1 + d_2 + R) - \frac{3}{4}(d_1 - d_2)^2\right)\frac{(d_1 + d_2 - 2R)^2}{4R} \tag{147}$$

so $|V_{12}| = \frac{1}{12}\pi(2d + R)(R - d)^2$. Hence

$$F_1 = \Phi_0(N) - \frac{R}{d}(\Phi_0(N) - \Phi_0(N_1) - \Phi_0(N_2)) \tag{148}$$

This differs from the exact result due to the term proportional to R. Tarazona and Rosenfeld suggest that the natural way to correct the functional is to introduce a correction to Φ_1 involving two-body contributions. Generalizing the expression for Φ_1, they propose that the correction be written as

$$F_2 = -\int d\mathbf{r}\,\psi(\eta(\mathbf{r}))\int ds_1 ds_2\,\rho(\mathbf{r} + s_1)w(s_1)\rho(\mathbf{r} + s_2)w(s_2)P(\mathbf{s}_1, \mathbf{s}_2) \tag{149}$$

A simple calculation, given in Appendix B, gives

$$F_2 = 4\pi\frac{d}{2}\left(\frac{1}{4\pi(d/2)^2}\right)^2\int d\mathbf{r}\,\psi''_0(\eta(\mathbf{r}))$$

$$\times \int ds_1 ds_2\,\rho(\mathbf{r} + s_1)w(s_1)\rho(\mathbf{r} + s_2)w(s_2)\left(\frac{d^2}{4} - \mathbf{s}_1 \cdot \mathbf{s}_2\right) \tag{150}$$

$$= \frac{1}{2\pi d}\int d\mathbf{r}\,\psi''_0(\eta(\mathbf{r}))(s^2(\mathbf{r}) - v^2(\mathbf{r}))$$

which is the same as the second contribution to the Rosenfeld functional.

So far, this exercise has just resulted in a somewhat different derivation of the Rosenfeld functional. Tarazona and Rosenfeld went on to consider the contributions of cavities formed by the intersection of three spherical cavities. They showed that the combination $\Phi_1 + \Phi_2$ does not give the correct result and thus motivates the inclusion of a three-body term, Φ_3. They discussed its properties, but an explicit form was only given later by Tarazona [83]. The result is expressed in terms of a tensor density measure,

$$T_{ij}(\mathbf{r}_1) = \int \rho(\mathbf{r}_2)\frac{r_{12,i}r_{12,j}}{r_{12}^2}\delta\left(\frac{d}{2} - r_{12}\right)dr_2 \tag{151}$$

as

$$\Phi_3 = \frac{3}{16\pi(1-\eta)^2}(\mathbf{v}\cdot\mathbf{T}\cdot\mathbf{v}-sv^2-\text{Tr}(\mathbf{T}^3)+s\,\text{Tr}(\mathbf{T}^2)) \tag{152}$$

where $Tr(\mathbf{A})$ indicates the trace of the tensor $\mathbf{A}$. Following Roth et al. [74], this can also be written in a more revealing form by separating the tensor into its trace and trace-less parts as

$$T_{ij}(\mathbf{r}_1) = \frac{1}{3}s(\mathbf{r}_1)\delta_{ij} + U_{ij}(\mathbf{r}_1)$$

$$U_{ij}(\mathbf{r}_1) = \int \rho(\mathbf{r}_2)\left(\frac{r_{12,i}r_{12,j}}{r_{12}^2}-\delta_{ij}\right)\delta\left(\frac{d}{2}-r_{12}\right)dr_2 \tag{153}$$

in terms of which the functional becomes

$$\Phi_3 = \frac{1}{24\pi(1-\eta)^2}\left(s^3-3sv^2+\frac{9}{2}(\mathbf{v}\cdot\mathbf{U}\cdot\mathbf{v}-\text{Tr}(U^3))\right) \tag{154}$$

This is a more natural representation in the sense that in a uniform liquid, both $\mathbf{v}$ and $\mathbf{U}$ vanish. It is also interesting because connections can be made to some of the earlier models mentioned above. Obviously, taking $U = 0$ gives the original Rosenfeld functional. One might imagine trying to approximate the tensor using the simpler densities as, for example,

$$U_{ij} \sim A\frac{v_iv_j-\frac{1}{3}v^2\delta_{ij}}{s} \tag{155}$$

where the prefactor is undetermined and the constraint on forming the right-hand side is that it must be a traceless tensor that scales linearly with the density. Putting this into Φ_3 gives

$$\Phi_3 \sim \frac{s^3}{24\pi(1-\eta)^2}(1-3\xi^2+3A\xi^4-A^3\xi^6) \tag{156}$$

Taking $A = 1$, the term in brackets becomes $(1-\xi^2)^3$, which is the same as the empirical model proposed by Rosenfeld et al. [80]. Tarazona has observed that Eq. 152 could also be guessed by postulating the need for a tensor density, constructing the lowest-order expression that vanishes in one dimension and that agrees with the first two orders of the virial expansion of the DCF in three dimensions, and guessing the prefactor $(1-\eta)^{-2}$ based on an analogy with the first two parts of the free energy [82].

The Tarazona functional has several significant properties in regard to the solid phase [83]. First, and perhaps most importantly, it stabilizes the solid phase and makes a reasonable prediction for liquid–solid coexistence (see Table II).

TABLE II
Comparison of the Predictions of Various FMT DFTs for the Freezing of Hard Spheres to Data from Simulation[a]

Theory	EOS	$\bar{\eta}_{\text{liq}}$	$\bar{\eta}_{\text{sol}}$	P^*	L
RSLT[b]	PY	0.491	0.540	12.3	1.06
Tarazona[c]	PY	0.467	0.516	9.93	0.145
White Bear[c,d]	CS	0.489	0.536	11.3	0.132
MC[e]	—	0.494	0.545	11.7	0.126

[a]Given are the liquid, $\bar{\eta}_{\text{liq}}$, and solid, $\bar{\eta}_{\text{sol}}$, packing fractions, the reduced pressure $P^* = \beta P d^3$ and the Lindemann parameter, L, at bulk coexistence. For each theory, the equation of state used for the fluid, Percus–Yevick(PY) or Carnahan–Starling (CS), is indicated. The Lindemann ratio for all three theories, calculated in the Gaussian approximation, is taken from Ref. 81.
[b]From Rosenfeld et al. [80].
[c]From Tarazona [82].
[d]From Roth et al. [74].
[e]From Hoover and Ree [57].

In fact, the description of the solid is rather good and it is the fact that the uniform liquid is described by the Percus–Yevick equation of state that is responsible for the difference from simulation. Tarazona also showed that the Lindemann parameter (i.e., the width of the Gaussians) is in good agreement with simulation at all densities and, in particular, that it vanishes as the density of the solid approaches close packing. Of the older, liquid-based DFTs, only the GELA had shown similar behavior and then only for the FCC phase. One of the virtues of FMT is that it contains enough geometrical information about hard spheres to show the expected divergences at close-packing. Groh and Mulder investigated the description of the solid phase without making the Gaussian approximation [28]. Their results confirm the improvement of the Tarazona model over earlier versions of FMT and also support the accuracy of the Gaussian approximation.

In a subsequent investigation of hard-sphere mixtures, Cuesta et al. related the ability of the FMT model to accurately describe these quasi-zero-dimensional systems to the low-density expansion of the three-body direct correlation function [84]. They showed that to described mixtures, a third-order tensor density measure must also be introduced. However, the theory is subject to instabilities leading to the possibility of infinitely negative free energies thus leaving its status in doubt. Their conclusion is that the natural extension of the Tarazona functional, Eq. (152), is preferable even though it is not formally as accurate as the more complex form.

E. The White-Bear Functional

One of the nice features of the various forms of FMT so far described is that the second functional derivative of the free energy functional with respect to density gives the Percus–Yevick DCF. However, this necessarily implies that the equation of state for the uniform fluid is the Percus–Yevick equation of state which is known to be inaccurate at moderate to high densities. There have therefore been several attempts to build in a more accurate equation of state. Roth, Evans, Lang, and Kahl proposed what is known as the "white-bear" functional based on a modified version of Rosenfeld's original derivation of FMT for mixtures [74] and a similar proposal for the monotonic system was made by Tarazona [82]. Recall that Eq. (124) above relates the pressure to the FMT ansatz and that Rosenfeld then goes on to eliminate the pressure using relations from SPT. Roth et al instead simply insert an empirical expression for the pressure written in terms of the density measures. The particular expression they use is the Mansoori–Carnahan–Starling–Leland (MCSL) expression which is a generalization of the Carnahan–Starling equation of state to mixtures [85]. The result is

$$\Phi_3 = \frac{s^3 - 3sv^2}{36\pi\eta^2(1-\eta)^2}\left(\eta + (1-\eta)^2\ln(1-\eta)\right) \tag{157}$$

They go on to propose that the factor $s^3 - 3sv^2$ be replaced by the expression occuring in Tarazona's functional to give the final white-bear functional,

$$\Phi_3 = \frac{s^3 - 3sv^2 + \frac{9}{2}(\mathbf{v}\cdot\mathbf{U}\cdot\mathbf{v} - \mathrm{Tr}(U^3))}{36\pi\eta^2(1-\eta)^2}\left(\eta + (1-\eta)^2\ln(1-\eta)\right) \tag{158}$$

Just as the uniform fluid is now described by the MCSL equation of state, this functional does not reproduce the Percus–Yevick DCF. A simple calculation gives the Percus–Yevick form,

$$c(r;d) = \left(a_0 + a_1\frac{r}{d} + a_3\left(\frac{r}{d}\right)^3\right)\Theta(d-r) \tag{159}$$

but with coefficients,

$$a_0 = -\frac{1 + 4\eta + 3\eta^2 - 2\eta^3}{(1-\eta)^4}$$

$$a_1 = \frac{2 - \eta + 14\eta^2 - 6\eta^3}{(1-\eta)^4} + \frac{2\ln(1-\eta)}{\eta} \tag{160}$$

$$a_3 = \frac{-3 + 10\eta - 15\eta^2 + 5\eta^3}{(1-\eta)^4} - \frac{3\ln(1-\eta)}{\eta}$$

When compared to computer simulation, this expression is shown to be more accurate, compared to simulations, than the Percus–Yevick approoximation [74]. Roth et al also show some small improvement over the Rosenfeld functional in the description of a fluid near a wall. A notable success is in giving much better values for liquid–solid coexistence than in the original theory of Tarazona (see Table II) due to the improved equation of state for the uniform liquid.

F. Calculating the Densities

In applications, one often needs to evaluate the density measures for particular parameterizations of the density (e.g., the Gaussian model of the solid) and/or for densities with particular symmetries (planar, spherical, . . .). The tensor density measures in particular appear complex, but their evaluation is not difficult when performed using a generating function. Consider in particular the functional

$$\tau(\mathbf{r}_1) = \frac{1}{2} \int d\mathbf{r}_2 \, \rho(\mathbf{r}_2) \left(\left(\frac{d}{2} \right)^2 - r_{12}^2 \right) \Theta \left(\frac{d}{2} - r_{12} \right) \tag{161}$$

It is easy to see that from this single scalar, all of the density measures up to the second-rank tensor follow:

$$\eta(\mathbf{r}) = \frac{4}{d} \frac{\partial}{\partial d} \tau(\mathbf{r})$$

$$s(\mathbf{r}) = 2 \frac{\partial}{\partial d} \eta(\mathbf{r})$$

$$v_i(\mathbf{r}) = -\partial_i \eta(\mathbf{r}) \tag{162}$$

$$T_{ij}(\mathbf{r}) = \frac{2}{d} \eta(\mathbf{r}) \delta_{ij} + \frac{2}{d} \partial_i \partial_j \tau(\mathbf{r})$$

Generating functionals that give all of these as well as even higher-order densities can be constructed by including additional factors of $\left(\frac{d}{2} \right) - r_{12}$ in the integral. In a planar geometry for which $\rho(\mathbf{r}) = \rho(z)$, one has

$$\tau(z) = \frac{\pi}{64} \int_{-d/2}^{d/2} \rho(z_1 + z)(d - 2z_1)^2 (d + 2z_1)^2 dz_1 \tag{163}$$

while in a spherical geometry, $\rho(\mathbf{r}) = \rho(r)$ and

$$\begin{aligned}
\tau(r) &= \frac{\pi}{4} \Theta \left(r - \frac{d}{2} \right) \frac{1}{r} \int_{-d/2}^{d/2} \rho(r_1 + r) \left(\left(\frac{d}{2} \right)^2 - r_1^2 \right)^2 (r_1 + r) dr_1 \\
&+ \frac{\pi}{4} \Theta \left(\frac{d}{2} - r \right) \frac{1}{r} \int_{d/2 - 2r}^{d/2} \rho(r_1 + r) \left(\left(\frac{d}{2} \right)^2 - r_1^2 \right)^2 (r_1 + r) dr_1 \\
&+ 2\pi \Theta \left(\frac{d}{2} - r \right) \int_0^{d/2 - r} \rho(r_1) \left(\left(\frac{d}{2} \right)^2 - r_1^2 - r^2 \right) r_1^2 dr_1
\end{aligned} \tag{164}$$

Finally, the Gaussian parameterization of the solid phase gives

$$\tau(\mathbf{r}_1) = x \sum_n \hat{\tau}(\mathbf{r}_1 - \mathbf{R}_n) \tag{165}$$

with

$$\hat{\tau}(\mathbf{r}_1) = \frac{1}{2}\left(\frac{\alpha}{\pi}\right)^{3/2} \int d\mathbf{r}_2 \exp(-\alpha r_2^2)\left(\left(\frac{d}{2}\right)^2 - r_{12}^2\right)\Theta\left(\frac{d}{2} - r_{12}\right)$$

$$= -\frac{x}{16\sqrt{\pi}\alpha^{3/2}r_1}\begin{pmatrix} 2(2-\alpha r_1 d + 2\alpha r_1^2)\exp\left(-\alpha\left(r_1 + \frac{d}{2}\right)^2\right) \\[2mm] -2(2+\alpha r_1 d + 2\alpha r_1^2)\exp\left(-\alpha\left(r_1 - \frac{d}{2}\right)^2\right) \\[2mm] +\sqrt{\alpha}r_1(6-\alpha d^2 + 4\alpha r_1^2)\sqrt{\pi}\left(erf\left(\sqrt{\alpha}\left(r_1 + \frac{d}{2}\right)\right)\right) \\[2mm] -erf\left(\sqrt{\alpha}\left(r_1 - \frac{d}{2}\right)\right) \end{pmatrix} \tag{166}$$

The representation in reciprocal space is much simpler:

$$\tau(\mathbf{r}) = x\bar{\rho}\sum_n \exp(i\mathbf{K}_n \cdot \mathbf{r})\exp(-K_n^2/4\alpha)\tilde{\tau}(\mathbf{K}_n) \tag{167}$$

with

$$\tilde{\tau}(\mathbf{K}) = -\pi\frac{d^2 K^2 \sin\frac{1}{2}Kd - 12\sin\frac{1}{2}Kd + 6Kd\cos\frac{1}{2}Kd}{K^5}$$

$$= \left(\frac{d}{2}\right)^5 \frac{4\pi}{105}\left(7j_0\left(\frac{Kd}{2}\right) + 10j_2\left(\frac{Kd}{2}\right) + 3j_4\left(\frac{Kd}{2}\right)\right) \tag{168}$$

where $j_n(x)$ is the nth-order spherical Bessel function.

G. Further Developments and Open Questions

The main virtues of FMT are (a) its internal consistency in that it does not require one DCF as input and imply another via the functional derivative of the model free energy functional and (b) its agreement with exact results in one and quasi-zero dimensions. The first property is shared by the MWDA and WDA effective liquid theories, but the second is particular to FMT.

Fundamental Measure Theory has been successfully used in a number of problems. In his original paper, Rosenfeld showed that it makes reasonable predictions for the three-body DCF [75]. Kierlik and Rosinberg found good results using the theory, together with a mean-field treatment of the long-ranged part of the potential, to describe adsorption of Lennard-Jones atoms at a wall [77, 79]. As discussed by Roth et al. [74], Rosenfeld's original FMT also gives a good description of hard spheres near a hard wall, and the white-bear functional is even better. Warshavsky and Song performed a remarkably difficult calculation of the free energy of the hard-sphere liquid–solid interface using FMT and found good results for the surface tension [86]. Similar results were found by Lutsko using a gradient theory constructed from FMT [87]. As discussed in the following section, FMT is widely used to model the short-ranged repulsion typical of simple fluids.

Despite its successes, there are still obstacles in the description of the solid phase in FMT. One problem is the development of multiple metastable solid states. Recall that in the Gaussian approximation, the value of the width of the Gaussians is determined by minimizing the free energy. Normally, one expects to find two minima: one at infinite width corresponding to the uniform liquid and one a some finite width corresponding to the solid. As shown by Lutsko [81], at high densities, the RSLT theory gives two nonzero minima for the FCC solid which, if taken seriously, would mean that there was a metastable solid phase. More seriously, the RSLT, Tarazona, and white-bear functionals all show two BCC solids for some density ranges, with the anomaly affecting the smallest range of densities in the case of the white-bear functional. Furthermore, while in all three theories one of the branches does show the correct vanishing of the Lindemann parameter at high densities, its behavior at intermediate densities is not monotonic, which is unphysical and not so different from what is found with the liquid-based theories. In fact, the only one of the three theories for which the *stable* branch of the free energy curve shows a vanishing Lindemann parameter is the Tarazona theory. The result is that the theory giving the best description of FCC melting (the white-bear) gives an unphysical picture of the BCC solid while the one giving the best description of the BCC phase (the Tarazona theory) gives a poor description of FCC melting and unphysical behavior for the BCC phase at intermediate densities.

Finally, it should be noted that various attempts have been made to extend FMT to other hard-core objects. Rosenfeld [88] has attempted to generalize the theory to hard, convex but nonspherical bodies while Cuesta and co-workers have constructed FMTs for parallel hard cubes [89, 90], hard-core lattice gases [91] and parallel hard cylinders [92]. Because these cases are all either very technical or somewhat artificial, the reader is referred to the literature for further details.

V. BEYOND HARD SPHERES

Fundamental Measure Theory gives a reasonable approximation to the DFT for hard spheres. Especially satisfying is the fact that it is constructed to agree with the exact DFT in several limiting cases and that it is closely related to the well-known Percus–Yevick theory (and SPT) in three dimensions. It is therefore natural to wonder whether the approach can be extended beyond hard objects. Rosenfeld's original development of FMT was closely tied to the geometry of hard spheres, and the later developments were intimately related to low-dimensional exact solutions that do not exist for long-ranged potentials. Thus the attempts that have been made are necessarily somewhat heuristic and, in fact, have met with mixed success [93, 94]. It is therefore the case that most applications to realistic potentials have relied on a separation of the contributions to the free energy into a hard-core contribution, modeled by FMT or one of the earlier DFTs for hard spheres, and a contribution due to the long-ranged interaction which is usually handled in a mean-field approximation. In this section, I discuss the basic construction of these models and give several illustrative examples of their application in practice.

It is important to specify what type of material is to be modeled. The primary distinction is between entities that interact via a potential with a strong short-ranged repulsion (such as simple liquids, colloids, proteins, ...) and soft matter, for which the interaction involves a weak short-ranged repulsion (such as some polymers). Indeed, in the extreme case that interparticle interaction is very *softly* repulsive (i.e., not divergent as $r \to 0$) and decays to zero at large distances, Likos et al. have shown that the Random Phase Approximation, $c(r) \sim -\beta v(r)$, is nearly exact at high densities [95]. A good quality, density-independent DCF allows for the construction of the free energy functional without further approximation. Thus, attention here will be focused on the first case of a strong repulsion for which such useful results are not available.

The underlying idea is that the DFT free energy functional is written as

$$F[\rho] = F_{id}[\rho] + F_{HS}([\rho]; d) + F_l[\rho] \tag{169}$$

where the first term on the right is the (exact) ideal-gas contribution, the second is the hard-sphere contribution (which depends on some hard-sphere diameter, d, which must be specified) and the last contribution accounts for the long-ranged interactions. This is motivated by the fact that the Mayer function for potentials with a strong short-ranged potential is quite similar to that for hard spheres, thus suggesting that the statistical mechanics of the latter might be viewed as "hard-spheres plus a correction." Furthermore, the analogy to thermodynamic perturbation theory is obvious. Since the basic physical intuition of separating the contributions into one modeled as a hard-core and a mean-field treatment of the attractive part is the same as that underlying the van der Waals equation of state, I will refer to these as van der Waals models (VdW) models. They are also often referred to in the literature as "mean-field models," although this is somewhat misleading because the different terms are actually treated at different levels of approximation. The two elements that remain to be specified are the hard-sphere diameter and the form of the long-ranged (or "tail") contribution.

An important concept is the division of the pair interaction potential, $v(r)$, into a short-ranged repulsive part, $v_0(r)$, and a long-ranged, attractive tail, $w(r)$. Taking the Lennard-Jones potential as an illustrative example,

$$v_{\mathrm{LJ}}(r) = 4\varepsilon\left(\left(\frac{\sigma}{r}\right)^{12} - \left(\frac{\sigma}{r}\right)^{6}\right) \tag{170}$$

there are two widely used choices which are taken form thermodynamic perturbation theory. The first is the Barker–Henderson (BH) division [21, 96],

$$
\begin{aligned}
v_0^{(\mathrm{BH})}(r) &= v_{\mathrm{LJ}}(r)\Theta(r_0 - r) \\
w^{(\mathrm{BH})}(r) &= v_{\mathrm{LJ}}(r)\Theta(r - r_0)
\end{aligned}
\tag{171}
$$

where the division is made at the point at which the potential is equal to zero, $v_{\mathrm{LJ}}(r_0) = 0$, giving $r_0 = \sigma$. The second is that used in the Weeks–Chandler–Anderson (WCA) theory [27, 97–99]

$$
\begin{aligned}
v_0^{(\mathrm{WCA})}(r) &= (v_{\mathrm{LJ}}(r) - v_{\mathrm{LJ}}(r_m))\Theta(r_m - r) \\
w^{(\mathrm{WCA})}(r) &= v_{\mathrm{LJ}}(r_m)\Theta(r_m - r) + v_{LJ}(r)\Theta(r - r_m)
\end{aligned}
\tag{172}
$$

where the division is made at the minimum of the potential, $v'_{\mathrm{LJ}}(r_m) = 0$, giving $r_m = 2^{1/6}\sigma$. The effective hard-sphere diameter is calculated using the $v_0(r)$ contribution. In the BH theory, it is given as

$$d_{\mathrm{BH}}(T) = \int_0^{r_0} (1 - \exp(-\beta v_0(r)))dr \tag{173}$$

which is chosen so as to optimize the representation of the short-ranged part of the potential by a hard-sphere interaction (see, e.g., the discussion in Ref. 21). Many other definitions occur in the literature on thermodynamic perturbation theory including that of the WCA theory [21, 97–99], the Lado theory [100] and the Mansoori–Canfield–Rasaiah–Stell theory [85]. While superior for use in thermodynamic perturbation theory in some situations (particularly at high density), these give density-dependent diameters and involve the PDF for hard spheres, both of which are problematic for DFT. A density-dependent diameter complicates the DFT because it adds an additional density dependence to the (already complicated) functionals. The need for the PDF is not too burdensome for liquids, but is more problematic for the solid phase (although there are available empirical [35, 101] and DFT-based [102–104] models) and is very difficult for interfacial problems involving, for example, transitions from liquid to solid. For these reasons, it is simpler to use the density-independent BH hard-sphere diameter or to use a diameter calculated for the bulk liquid. Other viable options would be the density-independent diameter obtained by matching the second virial coefficient of the hard-sphere and short-ranged interactions as used by Paricaud [105].

A.   Treatment of the Tail Contribution: Perturbative Theories

Returning to the exact formalism described in Section II.C, it is clear that the excess free energy functional is linear in the DCF. An exact separation into "short-ranged" and "long-ranged" contributions can therefore be made starting with the exact DCF by writing

$$c_2(\mathbf{r}_1, \mathbf{r}_2; [\rho]) = c_2^{\text{HS}}(\mathbf{r}_1, \mathbf{r}_2; d, [\rho]) + \Delta c_2(\mathbf{r}_1, \mathbf{r}_2; d, [\rho]) \qquad (174)$$

where the first term on the right is the exact DCF for hard spheres and the second term is defined by this expression. The excess free energy functional given in Eq. (44) will then also be a sum of a hard-sphere contribution and a part primarily related to long-ranged interactions,

$$\frac{1}{V}\beta F[\rho] = \beta F^{\text{HS}}(d, [\rho]) + \Delta \beta f_{\text{ex}}(\bar{\rho}_0; d) + \frac{\partial \Delta \beta f_{\text{ex}}(\bar{\rho}_0; d)}{\partial \bar{\rho}_0}\left(\bar{\rho}-\bar{\rho}_0\right)$$

$$-\frac{1}{V}\int_0^1 d\lambda \int_0^\lambda d\lambda' \int d\mathbf{r}_1 d\mathbf{r}_2 \Delta c_2(\mathbf{r}_1, \mathbf{r}_2; d, [(1-\lambda')\rho_0 + \lambda'\rho])$$

$$\times(\rho(\mathbf{r}_1)-\bar{\rho})(\rho(\mathbf{r}_2)-\bar{\rho}) \qquad (175)$$

where "Δ" always refers to the difference between the exact quantity and the corresponding hard-sphere quantity. The hard-sphere and long-ranged contributions can be handled differently. For the hard-sphere part, it is natural to use the FMT functional. For the remainder, all of the liquid-based theories discussed above remain viable options. However, since models like the MWDA, SCELA, and GELA work best for hard spheres and become less reliable as one moves to softer potentials, there is little incentive to use them for the long-ranged contribution. That leaves something like the simple perturbative theory of Ramakrishnan and Yussouff, Eq. (72), as the only viable option. Taking the reference state to be the uniform liquid and truncating at lowest order gives

$$\frac{1}{V}\beta F[\rho] = \beta F^{\text{HS}}(d, [\rho]) + \Delta \beta f_{\text{ex}}(\bar{\rho}_0; d) + \frac{\partial \Delta \beta f_{\text{ex}}(\bar{\rho}_0; d)}{\partial \bar{\rho}_0}(\bar{\rho}-\bar{\rho}_0)$$

$$-\frac{1}{2V}\int d\mathbf{r}_1 d\mathbf{r}_2 \Delta c_2(r_{12}; \bar{\rho}_0, d)(\rho(\mathbf{r}_1)-\bar{\rho}_0)(\rho(\mathbf{r}_2)-\bar{\rho}_0) \qquad (176)$$

which is an expression employed by Rosenfeld [106] using a DCF calculated from the MHNC and by Tang [107] using his first order mean spherical approximation.

For the case of a uniform solid or for a fluid near a wall, an obvious choice for the reference state exists (the average density of the solid and the bulk density of the fluid, respectively). However, this cannot be a "universal" functional because it obviously fails if the target state, $\rho(\mathbf{r}_1)$, is a uniform liquid with a different density than the reference state. This problem could be fixed if one kept the integration over density space and used, for example,

$$\frac{1}{V}\beta F[\rho] = \beta F^{\mathrm{HS}}(d, [\rho]) + \Delta\beta f_{\mathrm{ex}}(\bar{\rho}_0; d) + \frac{\partial \Delta\beta f_{\mathrm{ex}}(\bar{\rho}_0; d)}{\partial \rho_0}(\bar{\rho} - \bar{\rho}_0)$$

$$-\frac{1}{V}\int_0^1 d\lambda \int_0^\lambda d\lambda' \int d\mathbf{r}_1 d\mathbf{r}_2 \Delta c_2(r_{12}; d, (1-\lambda')\rho_0 + \lambda'\bar{\rho})$$

$$\times (\rho(\mathbf{r}_1) - \bar{\rho}_0)(\rho(\mathbf{r}_2) - \bar{\rho}_0) \tag{177}$$

which now gives the correct result when $\rho(\mathbf{r}) = \bar{\rho}$ for all values of the reference state. I am not aware of this having been used in practice. In fact, despite solving the problem of the uniform-liquid limit, this expression is still not satisfactory. Using the average bulk density for the reference density is reasonable in the examples given above, but for a problem such as the planar liquid–vapor interface, it makes little sense.

All of this suggest the use of a position-dependent reference density. The various models discussed in Section III.B.2 a can be used to construct local density approximations to the tail contribution. Some of these have been discussed in Ref. [42], where they were shown to give reasonable results for the description of liquid–vapor interfaces. However, the complexity of these models together with the fact that they are inconsistent in the sense discussed in Section III.B.1. e has led to other approaches being adopted.

B. Approaches Based on PDF

Once the free energy is broken into a hard-sphere contribution and the contribution due to the attractive part of the potential, it is natural to note the similarity to liquid-state perturbation theory where one has, at first order,

$$F(\rho; T) = F_{hs}(\rho d^3; T) + \frac{1}{2}\int w(r_{12})g_{hs}(r_{12}; d, T, \rho)d\mathbf{r}_1 d\mathbf{r}_2 \tag{178}$$

The idea is to use the last term on the right as a model for the treatment of the tail. Just as in the case of the DCF-based models, the problem is how to generalize this for an inhomogeneous system. Various proposals have been made such as that of Sokolowski and Fischer [108],

$$\Delta F[\rho] = \frac{1}{2}\int w(r_{12})\rho(\mathbf{r})_1\rho(\mathbf{r})_2 g_{hs}(r_{12}; d, T, \tilde{\rho}(\mathbf{r}_1, \mathbf{r}_2))d\mathbf{r}_1 d\mathbf{r}_2. \tag{179}$$

with

$$\tilde{\rho}(r_1, r_2) = \frac{1}{2}(\tilde{\rho}(r_1) + \tilde{\rho}(r_2))$$

$$\tilde{\rho}(r_1) = \frac{3}{4\pi R^3} \int \Theta(R-r)\rho(r)dr \tag{180}$$

where R is chosen to be some physically reasonable value on the order of the length scale of the potential. Sokolowski and Fischer found this to give good results for liquid–vapor coexistence and for the qualitative behavior of the liquid–wall interface. Wadewitz and Winkelmann applied a variant of the model to the calculation of Lennard-Jones surface tension and found reasonable results with the quality deteriorating with decreasing range of the potential [109]. A similar theory was also studied by Tang, Scriven, and Davis [110].

The advantage of these approximations is that one makes use of the well-developed machinery of thermodynamic perturbation theory and thereby assures a reasonable equation of state, at least for the bulk liquid. The drawback is that there is little relation to the DCF-based formalism of exact DFT. Furthermore, the dominant contribution to the DCF implied by Eq. (179) is likely to be $w(r_{12})g_{hs}(r_{12}; d, T, \tilde{\rho}(\mathbf{r}_1, \mathbf{r}_2))$, which, in a dense fluid, will exhibit strong oscillations coming from the hard-sphere PDF whereas such oscillations are not seen in simulation (see, e.g., Fig. 2 below).

C. Mean-Field Theories

Some of the inconsistencies of the effective-liquid approach to the tail contribution are avoided if Δc_2 is independent of density. The low-density limit of the DCF for any system (no matter what the density distribution) is

$$\lim_{\bar{\rho} \to 0} \Delta c_2(\mathbf{r}_1, \mathbf{r}_2; d, [\rho]) = -(1-\exp(-\beta v(r))) + \Theta(d-r) \tag{181}$$

which can be used as the desired approximation. It is more common, though, to break up the potential into a short-ranged repulsion and a long ranged attraction. Then, noting that the hard-sphere part of the DCF is supposed to model the short-ranged attraction, one has that $\exp(-\beta\, v_0(r)) \sim \Theta(r-d)$ so that

$$\lim_{\bar{\rho} \to 0} \Delta c_2(\mathbf{r}_1, \mathbf{r}_2; d, [\rho]) = -(1-\exp(-\beta v_0(r))\exp(-\beta w(r))) + \Theta(d-r)$$

$$\simeq \Theta(r-d)(\exp(-\beta w(r))-1) \tag{182}$$

At high temperatures, this gives the so-called Mean Spherical Approximation

$$\lim_{\beta \to 0}\lim_{\bar{\rho} \to 0} \Delta c_2(\mathbf{r}_1, \mathbf{r}_2; d, [\rho]) \simeq -\Theta(r-d)\beta w(r_{12}) \tag{183}$$

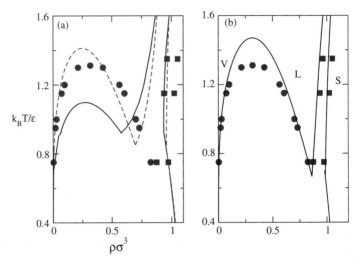

Figure 1. The phase diagram for the Lennard-Jones vapor (V), liquid (L), and FCC solid (S) calculated using different approximations. In all cases, the hard-sphere contribution is calculated using the white-bear FMT functional. (**a**) The tail contribution is calculated using Eq. 184 with both the BH (full line) and WCA (broken line) expressions for the long-ranged part of the potential. [Equation (173) was used to calculate the hard-sphere diameter in both cases.] The simulation data of Verlet and Levesque [113] and of Hansen and Verlet [114] are shown as symbols. (**b**) The result of using Eq. (186) with $w = 0$ and the effective hard-sphere diameter and perturbation theory of Ree et al. [115, 116].

For reasons discussed below, the preceding step function is often dropped and the approximation $\Delta c_2 \simeq -\beta w(r_{12})$ is used instead. It is interesting that besides being the expected result in the low-density, high-temperature limit, there is also evidence that this approximation becomes exact at high densities, at least in two dimensions [111]. This gives the mean-field result

$$F_l[\rho] = \frac{1}{2} \int \rho(\mathbf{r}_1)\rho(\mathbf{r}_2)w(r_{12})d\mathbf{r}_1 d\mathbf{r}_2 \qquad (184)$$

which has been used since the 1970's. Figure 1 shows the coexistence curves calculated using this model for the two choices of the tail function and the data from simulations. The liquid–vapor coexistence curve calculated using $w^{BH}(r)$ is significantly below the data while that calculated using $w^{WCA}(r)$ is above it. The difference is primarily due to the extension of the tail function into the core in the WCA model [compare Eq. (171) to Eq. (172)]. For the uniform fluid, one has that

$$\frac{1}{V}F_l^{(WCA)}(\bar{\rho}) = 2\pi\bar{\rho}^2 \int_0^\infty w^{(WCA)}(r)dr = \frac{2\pi}{3}\bar{\rho}^2 v_{LJ}(r_m)d^3 + 2\pi\bar{\rho}^2 \int_d^\infty w^{WCA}(r)dr \qquad (185)$$

The first term on the right is due to the extension of the "tail" into the core, while the second term is not very different from the full contribution of the BH tail. This is the reason for extending the tail contribution into the core: At least for a Lennard-Jones fluid, it appears to move the bulk liquid–vapor part of the phase diagram in the direction of the simulation results. Because the contribution of the tail function inside the core is of little importance in thermodynamic perturbation theory, the importance of this contribution to the VdW model is somewhat disturbing. In fact, this effect has been exploited by Curtin and Ashcroft [112] to give a better model for the LJ phase diagram by replacing $w^{(\mathrm{WCA})}(r)$ by $w^{(\mathrm{CA})}(r) = w^{(\mathrm{WCA})}(r)\Theta(r-r_*)$ with r_* chosen to be half the FCC nearest-neighbor distance. The rationale behind this choice is unclear except that it improves agreement of the phase diagram with the data. This model was subsequently used by Ohensorge, Löwen, and Wagner [59, 60] in an investigation of the liquid–solid interface near the triple point.

It is clear from Fig. 1 that the simple mean-field treatment of the tail is not very accurate, even for the liquid–vapor coexistence curve. Because the equation of state for the fluid can be relatively accurately calculated by other means (e.g., thermodynamic perturbation theory), this suggests introducing a correction based on this knowledge. For example, Lu, Evans, and Telo da Gamma [117] used a mean-field model but adjusted the hard-sphere diameter so as to reproduce the empirical liquid–vapor coexistence curve. For solids, a particularly simple model can be formulated in the context of the Gaussian density parameterization [Eq. (1.54)] where, for an inhomogeneous system, one allows the parameters, the average density $\bar{\rho} = x\bar{\rho}_{\mathrm{latt}}$, and the width of the Gaussians, α, to depend on position. Then, a very simple model giving the desired equation of state for the liquid is

$$F_l[\rho] = \frac{1}{2} \int \rho(\mathbf{r}_1)\rho(\mathbf{r}_2)w(r_{12})d\mathbf{r}_1 d\mathbf{r}_2 + \int \Delta f(\bar{\rho}(\mathbf{r}))d\mathbf{r} \qquad (186)$$

where $\Delta f(\rho)$ is the correction to the VdW model needed to give the known equation of state for the bulk liquid. An even simpler approximation drops the first term on the right altogether and models the tail solely through the effective liquid term. Both Curtin [58] and Lutsko and Nicolis [118] used this approximation to calculate the bulk equation of state for both the liquid and the solid. The phase diagram for the Lennard-Jones system is shown in Fig. 1. As is typical of all mean-field approaches, the equation of state (calculated using thermodynamic perturbation theory) is not accurate near the critical point. Otherwise, the predicted phase diagram is quantitatively very good. Similar results were obtained for the ten Wolde–Frenkel potential model for globular proteins [118]. The reason this works so well for the FCC solid is that the coordination in the solid and in the dense liquid is very similar so that the local environment of the atoms is not too different in the two systems. This approximation would not be expected to work for a BCC solid where something like the first term on the right in Eq. (186) would be necessary to account for the different structure.

A more elaborate modification of the VdW model which is closer to the spirit of FMT has recently been proposed [42] based on the observation that the simple VdW model implies a DCF in the uniform fluid of the form

$$c_2^{\text{VdW}}(r_{12}; \bar{\rho}) = c_2^{\text{FMT}}(r_{12}; \bar{\rho}, d) - \beta w(r_{12}) \tag{187}$$

provided that the hard-sphere diameter does not depend on the density. Comparison of this model to the DCF as determined by computer simulation shows that, at least in the dense fluid, most of the difference lies in the core region and, in fact, that the difference between the observed correlation function and the hard-sphere contribution is roughly linear in r. Motivated by the work of Tang [119], which primarily involves a complicated estimate of the DCF in the core region, this suggested that one might correct the correlations in the core by adding a linear contribution,

$$c_2^{\text{VdW}}(r_{12}; \bar{\rho}) = c_2^{\text{FMT}}(r_{12}; \bar{\rho}, d) + \Theta(d-r)\left(a_0(\bar{\rho}) + a_1(\bar{\rho})\frac{r}{d}\right) - \beta w(r_{12}) \tag{188}$$

with intercept and slope chosen so that the resulting DCF is continuous and gives some known equation of state via the compressibility equation [120],

$$-\beta w(d_+) = c_2^{\text{FMT}}(d_-; \bar{\rho}, d) + (a_0(\bar{\rho}) + a_1(\bar{\rho})) - \beta w(d_-)$$

$$f_{\text{ex}}(\bar{\rho}) = f_{\text{ex}}^{\text{HS}}(\bar{\rho}) - 4\pi d^3 \int_0^1 (1-\lambda)\left(\frac{1}{3}a_0(\lambda\bar{\rho}) + \frac{1}{4}a_1(\lambda\bar{\rho})\right) + 2\pi\bar{\rho}^2 \int_0^\infty w(r)dr$$

$$\tag{189}$$

where $f_{\text{ex}}(\bar{\rho})$ is the (excess part of the) desired equation of state for the uniform liquid and where the notation $d_\pm$ indicates the quantity evaluated at the hard-sphere diameter as it is approached from r greater than $(+)$ or less than $(-)$ d. In the spirit of the older liquid-based theories that often required the DCF of the uniform liquid as input, the equation of state is an input of the present theory and so must be determined, for example, from thermodynamic perturbation theory, the First-Order Mean Spherical Approximation of Tang [119], simulation, liquid-state integral equations, or some other source. As shown in Fig. 2, the improvement in the description of the DCF can be significant.

The second ingredient of the improved VdW model is to modify the free energy functional so as to be consistent with the modified DCF using the ideas of FMT. Following the approach of Kierlik and Rosinberg [77], who started with Rosenfeld's ansatz and demanded that it reproduce the Percus–Yevick DCF, one can do the same but demand that the desired core correction be obtained. The details are given in Ref. 42 and are straightforward, leading to

$$F[\rho] = F_{\text{id}}[\rho] + F_{\text{HS}}([\rho]; d) + F_{\text{core}}([\rho]; d) + F_l[\rho] \tag{190}$$

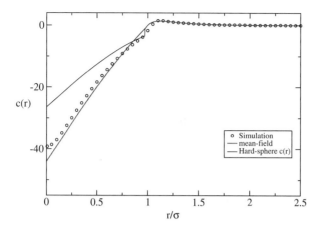

Figure 2. The DCF for a Lennard-Jones fluid at $T^* = 0.72$ and $\rho\sigma^3 = 0.85$ as determined from the model with the linear core correction, lower line, and the simulation data of Llano-Restrepo and Chapman [121]. The hard-sphere contribution to the DCF is shown as the upper line. From Ref. [120], where further details can be found.

with

$$F_{\text{core}}([\rho]; d) = \int \Phi_{\text{core}}(\mathbf{r}; [\rho]; d)d\mathbf{r}$$

$$\Phi_{\text{core}}(\mathbf{r}; [\rho]; d) = \frac{1}{\pi d^2}j_1(\eta(\mathbf{r}))s(\mathbf{r}) + \frac{1}{\pi d}j_2(\eta(\mathbf{r}))(s^2(\mathbf{r}) - v^2(\mathbf{r}))$$

$$+ \frac{1}{\pi}j_3(\eta(\mathbf{r}))s(\mathbf{r})(s^2(\mathbf{r}) - 3v^2(\mathbf{r})) \tag{191}$$

The quantities $\eta(\mathbf{r})$, and so on, are the density measures from FMT. Defining

$$\beta\phi_{\text{core}}(\bar{\rho}) = \frac{1}{\bar{\rho}}\left(f_{\text{ex}}(\bar{\rho}) - f_{\text{HS}}(\bar{\rho}) - \frac{1}{V}F_l(\bar{\rho})\right) \tag{192}$$

the functions $j_n(\eta)$ are given by

$$j_1(\eta) = \beta\phi_{\text{core}}(\bar{\rho}) + 3\eta(a_0(\bar{\rho}) + a_1(\bar{\rho})) + 72\eta^2 j_3(\eta)$$

$$j_2(\eta) = -\frac{1}{2}(a_0(\bar{\rho}) + a_1(\bar{\rho})) - 18\eta j_3(\eta)$$

$$j_3(\eta) = \frac{1}{36\eta^2}\left[\frac{1}{2}\left(\beta\phi_{\text{core}}(\bar{\rho}) - \bar{p}\frac{\partial}{\partial\bar{p}}\beta\phi(\bar{\rho}) - \beta\phi_{\text{core}}(0)\right)\right.$$

$$\left. + 3\int_0^\eta \eta\frac{d}{d\eta}c_{\text{HS}}(d_-; \bar{\rho}; d)d\eta\right] \tag{193}$$

where $\bar{\rho} = \bar{\rho}(\eta) = 6\eta/(\pi d^3)$.

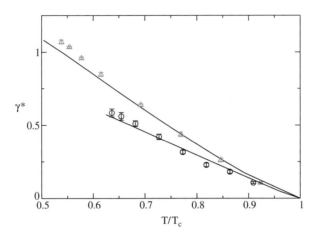

Figure 3. The surface tension for a Lennard-Jones fluid with a cutoff of 6σ (upper curve), and 2.5σ (lower curve). The solid lines are the result of the DFT calculations [42] and the data are from Grosfils and Lutsko [122]. The temperatures are scaled to the critical temperature while the dimensionless surface tension is $\gamma* = \gamma\sigma^2/\varepsilon$.

Figure 3 shows the excess free energy per unit area (the surface tension) for the planar Lennard-Jones vapor–fluid interface calculated using this model for different values of the cutoff of the potential and as determined by simulation [42]. The surface tension is very sensitive to the range of the potential, and this property can be used to test the robustness of a theory. The results are in good agreement with simulation, with most of the difference being due to inaccuracies in the equation of state at the small cutoff. For DFT to be useful, it must be transferable which is to say a given model must give reasonable results for a variety of potentials. The variation of surface tension with cutoff is one such test. As another illustration, the surface tension for the potential

$$v(r) = v_{\mathrm{HS}}(r, d) + \frac{4\varepsilon}{\alpha^2}\left(\left(\left(\frac{r}{\sigma}\right)^2 - d^2\right)^{-6} - \alpha\left(\left(\frac{r}{\sigma}\right)^2 - d^2\right)^{-3}\right) \qquad (194)$$

has also been calculated and compared to simulation. For $d = 0$ and $\alpha = 1$, this reduces to the Lennard-Jones potential; while for $d = \sigma$ and $\alpha = 50$, it is the interaction used by ten Wolde and Frenkel to model globular proteins [123]. Figure 4 shows the surface tension for various parameters, all with a cutoff of 2.8σ. In this case, the required equation of state is calculated using first-order thermodynamic perturbation theory, and the results are again seen to be in good agreement with the theory.

Figures 5 and 6 show the predicted density distribution for a Lennard-Jones fluid near a hard wall, along with the results of Grand Canonical Monte Carlo

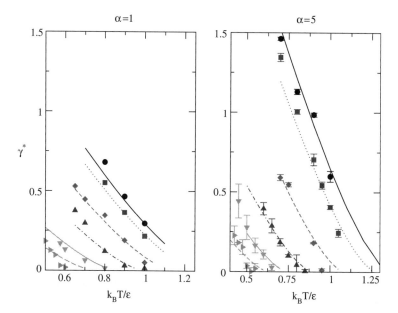

Figure 4. The surface tension for the potential given in Eq. (94) with a cutoff of 2.8σ, for different values of the hard-core diameter and for two different values of α. The points are from simulation with $d = 0.0$ (circles), 0.2(squares), 0.4 (diamonds), 0.6 (triangles), 0.8 (triangles down), and 1.0 (triangles right). The lines are from the DFT calculations [124].

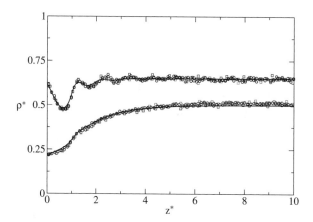

Figure 5. Structure of the Lennard-Jones fluid near a hard wall as determined from simulation (symbols) and the theory (lines). The simulations come from two runs each using cells with aspect ratio $1 \times 1 \times 2$ (circles), and $1 \times 1 \times 4$ (squares). The upper curve and data are for a chemical potential corresponding to bulk density $\rho^* = \rho\sigma^3 = 0.65$ and the lower curve for density $\rho^* = 0.50$. From Lutsko [42].

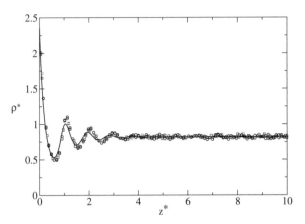

Figure 6.   Same as Fig. 5 except that the bulk density is $\rho^* = 0.85$. From Lutsko [42].

simulations as reported in Ref. 42. Again, the quantitative agreement is very good. Furthermore, it is significant that, because of its self-consistent structure, this relatively simple theory satisfies the exact sum rule that the density at the wall, $\rho(0)$, is the pressure in the bulk liquid, far from the wall, divided by the temperature (see Refs. 125, 126 and Appendix C). Physically, this is just a manifestation of the fact that the pressure in an equilibrium system must be uniform together with the fact that, when interacting with the wall, the particles behave like an

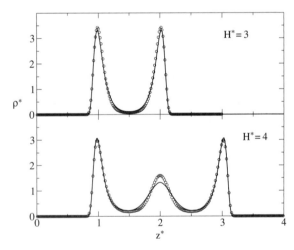

Figure 7.   Comparison of the density distribution of a Lennard-Jones fluid within slit pores of size $H^* = 3$ and $H^* = 4$ as calculated from the theory (lines) and as determined from simulation (symbols). From Lutsko [42].

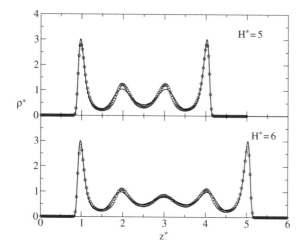

Figure 8. Comparison of the density distribution of a Lennard-Jones fluid within slit pores of size $H^* = 5$ and $H^* = 6$ as calculated from the theory (lines) and as determined from simulation (symbols). From Lutsko [42].

ideal gas. A final illustration is given in Figs. 7 and 8, which illustrate the predicted and observed density distribution of a Lennard-Jones fluid in a slit pore. A slit pore consists of two walls separated by some distance, H. The interaction between the fluid atoms and the walls was based on Steele's model [127, 128] of the average interaction of an atom with a 100 plane of an FCC solid, in which the potential felt by an atom a perpendicular distance z from the wall is

$$V_{\text{wall}}(z) = 2\pi\varepsilon \left(\frac{2}{5} \left(\frac{\sigma}{z} \right)^{10} - \left(\frac{\sigma}{z} \right)^4 - \frac{\sqrt{2}}{3\left(\frac{z}{\sigma} + 0.61/\sqrt{2} \right)^3} \right) \qquad (195)$$

The chemical potential was set to a value that corresponds to the bulk density $0.5925/\sigma^3$, the temperature was $k_B T = 1.2\epsilon$, and the intermolecular potential was cut off at a distance of 6σ. Again, the predicted density distribution is in good agreement with the simulations.

There have been many other applications of the general idea of writing the free energy as a sum of a hard-sphere contribution and of a mean-field treatment of the long-ranged part of the interaction. As just one example, Archer, Pini, Evans, and Reatto [129] have recently made an interesting study of colloidal fluids with competing interactions. In their work, DFT is compared to a sophisticated form of liquid-state theory and is found to predict interesting behavior (banding) in regions where the liquid-state theory has no solution.

VI. EXTENSIONS AWAY FROM EQUILIBRIUM

It seems intuitively obvious that if the local density of a system is perturbed, by applying some field, and then released, by removing the field, its relaxation will somehow be governed by the free energy surface $F[\rho]$. This idea, in less general forms, is quite old and obviously related to early work by Cahn as well as to the time-dependent Ginzburg–Landau model. In recent years, the idea has been developed into a set of techniques that combine the detailed free energy models developed in equilibrium DFT with methods of non-equilibrium statistical mechanics so as to allow for the description of dynamical transitions in complex systems. In the following, attention will mostly be given to the developments called Dynamical Density Functional Theory, with some discussion given also to the recent introduction of energy-surface methods for mapping out transition pathways.

A. Dynamical DFT

Density Functional Theory as described so far is a theory concerning equilibrium systems; and, as such, dynamics plays no role. Dynamical Density Functional Theory (DDFT), is an attempt to extend the ideas of DFT to dynamical properties. Intuitively, one expects that the free energy functionals used in DFT would play some role in determining the dynamics for systems out of equilibrium. For example, consider diffusion where number is conserved and so the density must obey a conservation law of the form

$$\frac{\partial \rho(\mathbf{r}, t)}{\partial t} = \mathbf{V} \cdot \mathbf{J}(\mathbf{r}, t) \tag{196}$$

where $\mathbf{J}$ is the number current. Assuming linear response and local equilibrium, the theory of non-equilibrium thermodynamics tells us that the thermodynamic force driving diffusion is the gradient in the local chemical potential [130] so, near equilibrium, one expects the current to have the form $\mathbf{J} = \mathbf{L} \cdot \mathbf{V}\mu(\mathbf{r}, t)$ where the tensor $\mathbf{L}$ is a transport coefficient related to the diffusion constant. If the system is in local equilibrium, the local chemical potential should be given by the Euler–Lagrange equation at each point so that the diffusion law becomes, in the simplest case of an isotropic system,

$$\frac{\partial \rho(\mathbf{r}, t)}{\partial t} = \mathbf{V} \cdot (LV\mu(\mathbf{r}, t)) = \mathbf{V} \cdot \left(LV \frac{\delta F[\rho]}{\delta \rho(\mathbf{r}, t)} \right) \tag{197}$$

Practical calculations based on nontrivial free energy models go back at least to Cahn [131]. A similar development for nonconserved variables, and including a fluctuating force, is called the Time-Dependent Ginzburg–Landau Theory and is a standard tool in the investigation of dynamical critical phenomena [132].

Notice that for a low-density system, $F[\rho] \sim F_{id}[\rho]$ giving

$$\frac{\partial \rho(\mathbf{r}, t)}{\partial t} \sim \mathbf{V} \cdot (L\rho(\mathbf{r}, t)^{-1} \mathbf{V}\rho(\mathbf{r}, t)) \tag{198}$$

which is the diffusion equation with a diffusion coefficient $D = L/\rho$. The diffusion coefficient diverges as the density goes to zero because the motion of each atom becomes ballistic—not diffusive—in the zero density limit. For mixtures, the limit of the density of *one* component going to zero does not produce ballistic motion since the finite component—the bath—still produces diffusive behavior. For the case of two components, with the tracer component, ρ_t, at vanishing low density and the bath component at low, spatially uniform, finite density ρ_b, where one has $\rho(\mathbf{r}, t)^{-1} = (\rho_t(\mathbf{r}, t) + \rho_b)^{-1} \sim \rho_b^{-1}$, the transport law becomes

$$\frac{\partial \rho_t(\mathbf{r}, t)}{\partial t} \sim \mathbf{V} \cdot (L\rho_b^{-1} \mathbf{V}\rho_t(\mathbf{r}, t)) \tag{199}$$

Now, if one supposes that the bath can be treated as a passive background, serving only to damp the dynamics of the tracers, and that the tracers can be treated as being in local equilibrium, then one might expect that for finite tracer density, Eq. (197) could be applicable with $F[\rho]$ being an effective free energy for the tracers. However, to be consistent with Eq. (199) in the low-density limit, one must assume that the transport coefficient is replaced by $L \rightarrow \Gamma\rho(\mathbf{r}, t)$, for some constant Γ, giving

$$\frac{\partial \rho}{\partial t} = \mathbf{V} \cdot \left(\Gamma\rho(\mathbf{r}, t) \mathbf{V} \frac{\delta F[\rho]}{\delta \rho(\mathbf{r}, t)} \right) \tag{200}$$

which is generally referred to as DDFT in the literature. This equation is assumed to describe the evolution of the density of a system subject to damping such as a colloidal fluid. Other systems for which the assumption of the passivity of the background is perhaps less plausible would include mixtures of similar species and the extreme case of self-diffusion.

The general form has a long history in the theory of non-equilibrium statistical mechanics. It is related to "Model B" in Hohenberg and Halperin's review of the theory of dynamical critical phenomena [132]. Some of the earliest uses of this and similar models, aside from that of Cahn, are Munakata [133, 134], Bagchi [135] and Dieterich [136]. The standard statistical mechanical approach to the derivation of equations such as Eq. (200) is by means of projection operators. For DDFT, the most complete analysis is probably that of Kawasaki [137]. An interesting derivation for granular systems was recently given by Tarazona and Marconi based on a multiscale expansion of the Enksog equation [138]. A very clear discussion of the relation between DDFT, the "extended" DDFT (i.e., DDFT including the local velocity), and conventional kinetic theory in the context of the glass transition can be found in Das [139]. Kirkpatrick and Wolynes [140] discuss the relation between equilibrium DFT and "extended" DDFT in the context of the glass transition. Chan and Finken discuss the formulation of an axiomatic DDFT in analogy to equilibrium DDFT [141]. Here, by way of illustration of the physical ideas involved, an intermediate route, based on recent work of Evans and Archer and of Marconi and Tarazona, will be followed wherein the introduction of the assumption of local equilibrium in otherwise exact balance laws leads to DDFT.

1.   Some ideas from Kinetic Theory

In order to discuss DDFT, it is first useful to review some basic ideas of kinetic theory. Consider a system of N particles (atoms, molecules, colloidal particles, ...) evolving under Newtonian dynamics. For simplicity, attention will be restricted here to entities having no internal structure so that they are fully described by their positions $\mathbf{q}_i$ and momenta $\mathbf{p}_i$, which together constitute the particle's phase $x_i = \mathbf{q}_i, \mathbf{p}_i$. The collection of all phases will be denoted $\Gamma_N = \{x_i\}_{i=1}^N$. The time-dependent N-body distribution $f(\Gamma_N; t)$ gives the probability to find the first particle with phase x_1, the second with phase x_2, and so on. It satisfies the Liouville equation,

$$\frac{\partial}{\partial t} f(\Gamma_N; t) + \sum_{i=1}^N \frac{\mathbf{p}_i}{m} \cdot \frac{\partial}{\partial \mathbf{q}_i} f(\Gamma_N; t) + \sum_{i=1}^N \frac{\partial}{\partial \mathbf{p}_i} \cdot \mathbf{F}_i f(\Gamma_N; t) = 0 \qquad (201)$$

where $\mathbf{F}_i$ is the total force acting on particle i. It will be the sum of any external forces, such as that due to an external field, and the internal forces due to the interparticle interaction potential so that for pair potentials it is

$$\mathbf{F}_i = \mathbf{F}_i^{\text{ext}} - \sum_{j \neq i} \frac{\partial}{\partial q_i} v(q_{ij}) \qquad (202)$$

Notice that the evolution of the phase function is deterministic and that stochasticity enters through the distribution of initial conditions. Defining the reduced distributions as

$$\rho_m(x_1, \ldots, x_m) = \frac{N!}{(N-m)!} \int f(\Gamma_N; t) dx_{m+1}, \ldots, dx_N \qquad (203)$$

and integrating the Liouville equation over particles $m+1, \ldots, N$ gives an equation for the m-body distribution containing a contribution due to the $(m+1)$-body distribution. This set of equations is known as the BBGKY hierarchy [142]. The first equation of the hierarchy is

$$\frac{\partial}{\partial t} \rho_1(x_1; t) + \frac{\mathbf{p}_1}{m} \cdot \frac{\partial}{\partial \mathbf{q}_1} \rho_1(x_1; t) + \frac{\partial}{\partial \mathbf{p}_1} \cdot \mathbf{F}_1^{\text{ext}} \rho_1(x_1; t) = \frac{\partial}{\partial \mathbf{p}_1} \cdot \int \frac{\partial v(q_{12})}{\partial \mathbf{q}_1} \rho_2(x_1, x_2; t) dx_2$$

$$(204)$$

In equilibrium, the distributions are time-independent and the velocities are always distributed as Maxwellians so that

$$
\begin{aligned}
\rho_1(x_1; t) &\rightarrow \rho(\mathbf{q}_1)\varphi(\mathbf{p}_1; \mathbf{u}, T) \\
\rho_2(x_1, x_2; t) &\rightarrow \rho(\mathbf{q}_1, \mathbf{q}_2)\varphi(\mathbf{p}_1; \mathbf{u}, T)\varphi(\mathbf{p}_2; \mathbf{u}, T)
\end{aligned}
\tag{205}
$$

where $\rho(\mathbf{q}_1)$ is the usual one-body density, $\rho(\mathbf{q}_1, \mathbf{q}_2)$ is the two-particle reduced distribution, which is related to the PDF as discussed in Section II, and the Maxwellian velocity distribution is

$$
\varphi(\mathbf{p}; \mathbf{u}, T) = \left(\frac{1}{2\pi m k_B T} \right)^{D/2} \exp\left(-\frac{(\mathbf{p}-\mathbf{u})^2}{2m k_B T} \right)
\tag{206}
$$

(The center of mass velocity $\mathbf{u}$ plays no role here, but we include it for the sake of a later discussion.) Substituting this into the equation for the one-body equation, multiplying through by $\mathbf{p}_1$ and integrating over $\mathbf{p}_1$ gives

$$
-k_B T \frac{\partial}{\partial \mathbf{q}_1} \rho(\mathbf{q}_1) + F_1^{\text{ext}} \rho(\mathbf{q}_1) = \int \frac{\partial v(q_{12})}{\partial \mathbf{q}_1} \rho(\mathbf{q}_1, \mathbf{q}_2) dx_2
\tag{207}
$$

In equilibrium, the density satisfies the Euler–Lagrange equation, Eq. (31), and using it in the left-hand side gives

$$
-\rho(\mathbf{q}_1) \frac{\partial}{\partial \mathbf{q}_1} \frac{\delta F_{\text{ex}}[\rho]}{\delta \rho(\mathbf{q}_1)} = \int \frac{\partial \beta v(q_{12})}{\partial \mathbf{q}_1} \rho(\mathbf{q}_1, \mathbf{q}_2) dx_2
\tag{208}
$$

which is the well-known first member of the Yvon–Born–Green hierarchy equation [21]. This exact, equilibrium relation plays a central role in several recent derivations of DDFT as described below.

2. A Simple Kinetic-Theory Approach

In general, the one-body density $\rho_1(x_1; t)$ depends on both position and velocity in nontrivial ways which are hard to calculate due to the coupled nature of the BBGKY hierarchy. However, integrating it over momenta gives the local density,

$$
\rho(\mathbf{r}; t) = \int \delta(\mathbf{r}-\mathbf{q}_1) \rho_1(x_1; t) dx_1
\tag{209}
$$

and this satisfies a simple continuity equation. Multiplying the first BBGKY equation by $\delta(\mathbf{r}-\mathbf{q}_1)$ and integrating over momentum gives

$$
\frac{\partial}{\partial t} \rho(\mathbf{r}; t) + \nabla \rho(\mathbf{r}; t) \mathbf{v}(\mathbf{r}; t) = 0
$$

where the local velocity is defined as

$$
\rho(\mathbf{r}; t) \mathbf{v}(\mathbf{r}; t) = \int \frac{1}{m} p_1 \delta(\mathbf{r}-\mathbf{q}_1) \rho_1(x_1; t) dx_1
\tag{210}
$$

The equation for the density has the expected form with the number current $\mathbf{J}(\mathbf{r}; t) = \rho(\mathbf{r}; t)\mathbf{v}(\mathbf{r}; t)$. A balance equation for the local velocity can be obtained in a similar way, by multiplying the first BBGKY equation by $\frac{\mathbf{p}_1}{m}\delta(\mathbf{r}-\mathbf{q}_1)$ and integrating to get

$$
m\frac{\partial}{\partial t}\rho(\mathbf{r}; t)\mathbf{v}(\mathbf{r}; t) + \nabla \cdot \int \frac{\mathbf{p}_1\mathbf{p}_1}{m}\delta(\mathbf{r}-\mathbf{q}_1)\rho_1(x_1; t)dx_1 - \mathbf{F}^{\text{ext}}(\mathbf{r}; t)\rho(\mathbf{r}; t)
$$
$$
= -\frac{1}{m}\int \delta(\mathbf{r}-\mathbf{q}_1)\frac{\partial v(q_{12})}{\partial \mathbf{q}_1}\rho_2(x_1, x_2; t)dx_1 dx_2 \tag{211}
$$

Changing integration variables from $\mathbf{p}_i$ to $\mathbf{p}'_i = \mathbf{p}_i - m\mathbf{v}(\mathbf{q}_i; t)$ and rearranging gives the exact balance equation

$$
\frac{\partial}{\partial t}\rho(\mathbf{r}; t)\mathbf{v}(\mathbf{r}; t) + \nabla \cdot \rho(\mathbf{r}; t)\mathbf{v}(\mathbf{r}; t)\mathbf{v}(\mathbf{r}; t)
$$
$$
+ \nabla \cdot \int \frac{\mathbf{p}'_1\mathbf{p}'_1}{m^2}\delta(\mathbf{r}-\mathbf{q}_1)\rho_1(x'_1; t)dx'_1 + \frac{1}{m}\int \delta(\mathbf{r}-\mathbf{q}_1)\frac{\partial v(q_{12})}{\partial \mathbf{q}_1}\rho_2(x_1, x_2; t)dx_1 dx_2
$$
$$
- \frac{1}{m}\mathbf{F}^{\text{ext}}(\mathbf{r}; t)\rho(\mathbf{r}; t) = 0 \tag{212}
$$

Separating out the trace and traceless part of the $\mathbf{p}'_1\mathbf{p}'_1$ term, noting that the potential term does not involve momenta, and using the continuity equation gives

$$
\rho(\mathbf{r}; t)\frac{\partial}{\partial t}\mathbf{v}(\mathbf{r}; t) + \rho(\mathbf{r}; t)\mathbf{v}(\mathbf{r}; t) \cdot \nabla\mathbf{v}(\mathbf{r}; t) + \nabla \cdot \frac{1}{m}\rho(\mathbf{r}; t)k_B T(\mathbf{r}; t)
$$
$$
+ \frac{1}{m}\int \delta(\mathbf{r}-\mathbf{q}_1)\frac{\partial v(q_{12})}{\partial \mathbf{q}_1}\rho_2(\mathbf{x}_1, \mathbf{x}_2; t)dx_1 dx_2 + \nabla \cdot \Pi^K - \frac{1}{m}\mathbf{F}^{\text{ext}}(\mathbf{r}; t)\rho(\mathbf{r}; t) = 0 \tag{213}
$$

with the local temperature defined as

$$
\frac{D}{2}\rho(\mathbf{r}; t)k_B T(\mathbf{r}; t) = \int \frac{(p_1 - m\mathbf{v}(\mathbf{q}_1; t))^2}{2m}\delta(\mathbf{r}-\mathbf{q}_1)\rho_1(x_1; t)dx_1 \tag{214}
$$

and where the kinetic part of the dissipative stress is

$$
\Pi^K_{ij} = \int \frac{\mathbf{p}'_{1i}\mathbf{p}'_{1j} - \frac{1}{D}p'^2_1\delta_{ij}}{m^2}\delta(\mathbf{r}-\mathbf{q}_1)\rho_1(x'_1; t)dx'_1 \tag{215}
$$

Similarly, a balance equation for the temperature is obtained by multiplying the first BBGKY equation by $\frac{1}{2m}p_1^2\delta(\mathbf{r}-\mathbf{q}_1)$ and integrating over momenta. After some manipulations, the result is

$$\frac{D}{2}\rho\frac{\partial}{\partial t}k_B T + \frac{D}{2}\rho\mathbf{v}\cdot\nabla k_B T + \Pi:\nabla\mathbf{v}$$

$$= -\nabla\cdot\int\frac{p_1'}{m}\frac{1}{2m}p_1'^2\delta(\mathbf{r}-\mathbf{q}_1)\rho_1(x_1;t)dx_1 - \int\frac{p_1'}{m}\delta(\mathbf{r}-\mathbf{q}_1)\cdot\frac{\partial v(q_{12})}{\partial\mathbf{q}_1}\rho_2(x_1,x_2;t)dx_2$$

$$(216)$$

So far, these equations are exact and therefore quite formal, requiring the one and two body distributions for closure. However, for an *isothermal* system, the equations for the density and velocity decouple from the temperature equation. Following Archer and Rauscher [143] and introducing the local equilibrium hypothesis for the two-body term, Eq. (208), in the velocity equation then gives

$$\frac{\partial}{\partial t}\rho(\mathbf{r};t) + \nabla\cdot\rho(\mathbf{r};t)\mathbf{v}(\mathbf{r};t) = 0$$

$$\frac{\partial}{\partial t}\mathbf{v}(\mathbf{r};t) + \mathbf{v}(\mathbf{r};t)\cdot\nabla\mathbf{v}(\mathbf{r};t) + \frac{1}{m}\nabla\frac{\delta F[\rho]}{\delta\rho(\mathbf{r};t)} + \rho(\mathbf{r};t)^{-1}\nabla\cdot\Pi - \frac{1}{m}\mathbf{F}^{\text{ext}}(\mathbf{r};t) = 0$$

$$(217)$$

To proceed further requires some sort of approximation for the dissipative stress. Most practical methods involve expressing it as an expansion in gradients of the density and velocity. One route is to use an approximate kinetic theory, such as the Boltzmann or Enskog equation, and the Chapman–Enskog expansion to obtain an analytic approximation to the gradient expansion of Π [142]. Alternatively, one can treat the term phenomenologically. In both cases, one could in principle keep the equation for the temperature as well, treating all difficult terms in the same way as Π. In all cases, the resulting equations have the interpretation of being the usual Navier–Stokes equations with a position-dependent pressure given by

$$m\rho(\mathbf{r};t)^{-1}\nabla p = \nabla\frac{\delta F[\rho]}{\delta\rho(\mathbf{r};t)} \qquad (218)$$

which suggests another interpretation. The Gibbs–Duhem equation for a multi-component system is

$$\sum_i N_i d\mu_i = -SdT + Vdp \qquad (219)$$

where N_i is the number of particles of species i and μ_i is the chemical potential for species i. Assuming local equilibrium, one finds for an isothermal, single-component system

$$\rho(\mathbf{r};t)d\mu(\mathbf{r};t) = dp(\mathbf{r};t) \Rightarrow \rho(\mathbf{r};t)\nabla\mu(\mathbf{r};t) = \nabla p(\mathbf{r};t)$$

$$\Rightarrow \rho(\mathbf{r};t)\nabla\frac{\delta F[\rho]}{\delta\rho(\mathbf{r};t)} = \nabla p(\mathbf{r};t) \qquad (220)$$

where the last line comes from the Euler–Lagrange equation and the assumption of local equilibrium. Thus, this generalized form of the Navier–Stokes equations can again be understood as the result of the use of an assumption of local equilibrium.

To make a connection to DDFT, which involves only the density, requires the additional approximation that the velocity responds much more quickly than does the density—that is, that the system is overdamped. In this case, for a given configuration of the density, the velocity quickly reaches a steady state driven by the gradient in the "pressure". In this state, one imagines that the velocity will be proportional to the driving force,

$$\mathbf{v}(\mathbf{r};t) = -\Gamma\left(\nabla\frac{\delta F[\rho]}{\delta\rho(\mathbf{r};t)} + \mathbf{F}^{\text{ext}}(\mathbf{r};t)\right) \qquad (221)$$

where Γ is a constant. [This is equivalent to looking neglecting the convective term in the second line of Eq. (217) and assuming that $\nabla\cdot\Pi \sim \Gamma^{-1}\mathbf{v}$, the same assumptions leading to Darcy's law. For example, this condition occurs naturally if the environment exerts a global friction as is the case in, e.g. Brownian dynamics [154]]. Then, the continuity equation gives

$$\frac{\partial}{\partial t}\rho(\mathbf{r};t) = \Gamma\nabla\cdot\rho(\mathbf{r};t)\left(\nabla\frac{\delta F[\rho]}{\delta\rho(\mathbf{r};t)} + \mathbf{F}^{\text{ext}}(\mathbf{r};t)\right) \qquad (222)$$

which is the expected result.

Just as in the case of DDFT, a more systematic approach to the derivation of this "extended DDFT" is based on the projection operator techniques of Zwanzig [144] and Mori and co-workers [145, 146]. However, this more general model is much older than DDFT and is widely used in Kinetic Theory today (see, e.g., Ref. 47). For a one-component liquid, the result is a generalized Langevin equation of the form of Eq. (217) with the addition of fluctuating forces. The construction of this and related models is outlined in Ref. 148 and the case of the pure fluid is given explicitly in Refs. 149 and 150. In this mesoscopic picture, Eq. (217) is understood as the result of averaging over the noise, yielding the same dynamics but with renormalized transport coefficients, as discussed more fully below.

3. Brownian Dynamics

One case in which the velocities of the particles may often be ignored is when the particles are in some sort of solution. In the event that motion is damped by the bath, the dynamics can be approximated as that of a collection of particles moving under the effect of their mutual interactions as well as a friction proportional to their velocities. Many derivations have been based on the ideas of Marconi and Tarazona [151] in which the starting point is a system described by this Brownian dynamics. The particles move according to the stochastic equations

$$m\frac{d^2\mathbf{q}_i}{dt^2} + \Gamma^{-1}\frac{d\mathbf{q}_i}{dt} = -\frac{\partial}{\partial \mathbf{q}_i}\left[\sum_{j \neq i} v(\mathbf{q}_{ij}) + \phi(\mathbf{q}_i)\right] + \eta_i(t) \qquad (223)$$

where the first term on the right is the sum of the interparticle force and the force due to interaction with a one-body potential while $\eta_i(t)$ is white noise representing interaction with the bath [152],

$$\langle \eta_i(t) \rangle = 0$$
$$\left\langle \eta_i(t)\eta_j(t') \right\rangle = 2k_B T \delta_{ij}\delta(t-t') \qquad (224)$$

The constant Γ is a measure of the friction due to the bath. In the limit of strong friction, $\Gamma^{-1} \gg 1$, the second-order time derivative can be ignored, giving a first-order equation of motion usually referred to as Brownian dynamics. Let $P(\mathbf{q}_1, \ldots, \mathbf{q}_N; t)$ be the probability to find particle 1 at position $\mathbf{q}_1$, and so on at time t. Then, the corresponding Fokker–Planck equation is [151, 152]

$$\frac{\partial}{\partial t}P(\mathbf{q}_1, \ldots, \mathbf{q}_N; t) = \Gamma \sum_{i=1}^{N} \frac{\partial}{\partial \mathbf{q}_i} \cdot \left\{ k_B T \frac{\partial}{\partial \mathbf{q}_i} + \sum_{j \neq i} \frac{\partial v(\mathbf{q}_{ij})}{\partial \mathbf{q}_i} \right\} P(\mathbf{q}_1, \ldots, \mathbf{q}_N; t)$$

$$(225)$$

Following Evans and Archer [153], we proceed as in the derivation of the BBGKY hierarchy above. The one-body density is

$$\rho(\mathbf{r}_1; t) = \left\langle \sum_i \delta(\mathbf{r}_1 - \mathbf{q}_i); t \right\rangle$$
$$= \int \sum_i \delta(\mathbf{r}_1 - \mathbf{q}_i) P(\mathbf{q}_1, \ldots, \mathbf{q}_N; t) d\mathbf{q}_1 \ldots d\mathbf{q}_N$$
$$= N \int P(\mathbf{r}_1, \mathbf{q}_2 \ldots, \mathbf{q}_N; t) d\mathbf{q}_2 \ldots d\mathbf{q}_N \qquad (226)$$

So that an equation for its time evolution can be obtained by integrating the Fokker–Planck equation to get

$$\frac{\partial}{\partial t}\rho(\mathbf{r}_1; t) = \Gamma \frac{\partial}{\partial \mathbf{r}_1}\left\{ k_B T \frac{\partial}{\partial \mathbf{r}_1}\rho(\mathbf{r}_1; t) + \int \frac{\partial v(\mathbf{r}_{12})}{\partial \mathbf{r}_1}\rho(\mathbf{r}_1, \mathbf{r}_2; t)d\mathbf{r}_2 \right\} \qquad (227)$$

where

$$\rho(\mathbf{r}_1, \mathbf{r}_2; t) = N(N-1)\int P(\mathbf{r}_1, \mathbf{r}_2, \mathbf{q}_3 \ldots, \mathbf{q}_N; t)d\mathbf{q}_3 \ldots d\mathbf{q}_N \qquad (228)$$

Assuming that the equilibrium configuration of the colloidal particles under the influence of a field will be the same as that of the particles in the absence of a bath,

Eq. (208) holds for the *equilibrium* one-body distribution. Introducing this here as an assumption of local equilibrium gives

$$
\frac{\partial}{\partial t}\rho(\mathbf{r};t) = \Gamma\nabla\cdot\left\{k_BT\nabla\rho(\mathbf{r};t) + \rho(\mathbf{r};t)\nabla\frac{\delta F_{\text{ex}}[\rho]}{\delta\rho(\mathbf{r};t)}\right\}
$$

$$
= \Gamma\nabla\cdot\left\{\rho(\mathbf{r};t)\nabla\frac{\delta F[\rho]}{\delta\rho(\mathbf{r};t)}\right\}
$$

(229)

which is the expected result.

4. A Note on Interpretation

There has been some discussion in the literature on whether or not a fluctuating force should be included in the DDFT equation [143, 151, 153, 155]. In fact, a careful examination of the derivations should give the answer in any particular case. The confusion is due to the fact that the same symbol is used by different workers for different objects leading, for example, to claims that some results make no sense [151]. The most fundamental level of description concerns the time evolution of the microscopic density, $\hat{\rho}(\mathbf{r}, t) = \sum_i\delta(\mathbf{r}-\mathbf{q}_i(t))$. This is just the Liouville equation. At a mesoscopic level, one projects out all except the slow variables associated with the local values of the conserved densities of number, momentum, and energy [142]. This results in the generalized Langevin equation, an exact equation, which certainly contains a noise term. Note that the density that occurs in this (exact) equation need not be the microscopic density $\hat{\rho}(\mathbf{r}, t)$ but is more commonly some coarse-grained mesoscopic density $\tilde{\rho}(\mathbf{r}, t)$. The presence of the noise term means that not all of the averaging has been done and so the local density appearing in the equations is not the fully ensemble-averaged quantity, $\rho(\mathbf{r}, t)$. The "free energy" appearing in these equations is also not, strictly speaking, the free energy functional of DFT. Typically, the "free energy functional" is actually a quantity of the form

$$
\exp(-\beta\mathcal{F}([\tilde{\rho}], [\phi]; t)) = \int \delta(\tilde{\rho}(\mathbf{r})-\hat{\rho}(\mathbf{r}))f_{\text{eq}}(q_1\ldots x_N; \phi)dx_1\ldots dx_N
$$

(230)

where $\hat{\rho}$ is the microscopic density function, $f_{\text{eq}}(q_1\ldots x_N; \phi)$ is the *equilibrium* distribution and the external field is ϕ. (Note that the equilibrium distribution appears here, rather than the non-equilibrium distribution, because of the definition of projection operators for which one chooses the equilibrium distribution as the measure in phase space. This is therefore *not* the result of a local equilibrium approximation.) The definition of the delta function is actually delicate and is perhaps best thought of in a course-grained sense. For example, space could be divided into (arbitrarily) small cells, $C^{(m)}$, of volume $V^{(m)}$ with centers at the points

$\mathbf{r}^{(m)}$ for $m = 1, \ldots M$, and the mesoscopic variables can be taken to be the densities in each cell,

$$\hat{\rho}^{(m)} = \frac{1}{V^{(m)}} \int_{C^{(m)}} \hat{\rho}(\mathbf{r}) d\mathbf{r}. \tag{231}$$

In this case, the precise meaning of the delta function is that it fixes the density in the mth cell to be some specified value, $\tilde{\rho}^{(m)}$,

$$\delta(\tilde{\rho}(\mathbf{r}) - \hat{\rho}(\mathbf{r})) \Rightarrow \prod_{m=1}^{M} \delta(\hat{\rho}^{(m)} - \tilde{\rho}^{(m)}) \tag{232}$$

Thus, the collection of values $\{\tilde{\rho}^{(m)}\}$ constitutes the "function" $\tilde{\rho}(\mathbf{r})$. Then, the actual free energy would be obtained by means of a further average,

$$\exp(-\beta F[\phi]) = \int_0^\infty d\tilde{\rho}^{(1)} \ldots \int_0^\infty d\tilde{\rho}^{(m)} \exp(-\beta \mathcal{F}([\tilde{\rho}], [\phi])) \delta\left(N - \sum_{m=1}^{M} \tilde{\rho}^{(m)} V^{(m)}\right) \tag{233}$$

and it is the quantity $F[\phi]$ with which contact is made with DFT. (Note that in this expression, it is assumed that the number of particles is fixed; the same arguments could be made in the grand canonical ensemble.) The constrained "free energy" $\mathcal{F}([\tilde{\rho}], [\phi])$ is the one that is discussed in field-theoretic approaches to statistical mechanics and is well-approximated via a mean-field model [156]. The true free energy $F[\phi]$ is a result of averaging over the coarse-grained density and includes renormalization effects as discussed by Reguerra and Reiss [157]. In terms of DDFT, the dynamical equations for the mesoscopic density $\tilde{\rho}(\mathbf{r}, t)$ would include a noise term,

$$\frac{\partial}{\partial t} \tilde{\rho}(\mathbf{r}; t) = \Gamma_0 \mathbf{\nabla} \cdot \left\{ \tilde{\rho}(\mathbf{r}; t) \mathbf{\nabla} \frac{\delta F([\tilde{\rho}], [\phi])}{\delta \tilde{\rho}(\mathbf{r}; t)} \right\} + \eta(\mathbf{r}; t) \tag{234}$$

The ensemble-averaged density is then the result of averaging this over the noise (i.e., the remaining degrees of freedom)

$$\frac{\partial}{\partial t} \langle \tilde{\rho}(\mathbf{r}; t) \rangle_\eta = \Gamma_0 \mathbf{\nabla} \cdot \left\{ \left\langle \tilde{\rho}(\mathbf{r}; t) \mathbf{\nabla} \frac{\delta F([\tilde{\rho}], [\phi])}{\delta \tilde{\rho}(\mathbf{r}; t)} \right\rangle_\eta \right\} \tag{235}$$

or

$$\frac{\partial}{\partial t} \rho(\mathbf{r}; t) \simeq \Gamma \mathbf{\nabla} \cdot \left\{ \rho(\mathbf{r}; t) \mathbf{\nabla} \frac{\delta F([\rho])}{\delta \rho(\mathbf{r}; t)} \right\} \tag{236}$$

where it is noted that typically, transport coefficients like Γ_0 are renormalized by the evaluated. (This supposes that the cells used in the coarse graining are sufficiently small that the noise-averaged coarse grained density and ensemble averaged density are the same.)

In the case of Brownian dynamics, the situation is completely analogous. In the derivation given above, the starting point—the Brownian dynamical equations—are the result of projecting out the degrees of freedom of a bath and so are already (an approximation to) the generalized Langevin equation. Statistical averages are then evaluated with respect to the noise and the one-body distribution is in fact $\rho(\mathbf{r}; t)$, the noise-averaged one-body density $\langle \tilde{\rho}(\mathbf{r}; t) \rangle_\eta$. Thus, one derives the equivalent of Eq. (235) and not Eq. (234). It might seem that there is some advantage in this approach, rather than the projection operator derivation, because one avoids coarse-graining. However, this ignores the fact that the concept of local equilibrium is itself based on the idea that the system can be divided into small volumes, each of which is in an equilibrium state, but with thermodynamic variables varying from volume to volume [130, 158]. The main advantage of the systematic approach, such as that of Kawasaki [137], over the heuristic invocation of local equilibrium is to make such assumptions explicit. However, in practical applications, the use of local equilibrium, particularly Eq. (208), may be the shortest route to a DDFT-like description.

5. Applications of DDFT

There have been many interesting uses of DDFT and here a sampling of the literature is given to illustrate the range of applications. Evans and Archer used it to study the kinetics of spinodal decomposition [153]. Archer, Hopkins, and Schmidt [159] used DDFT to calculate the van Hove dynamic correlation function for a simple fluid (the generalization of the static PDF giving the probability to find a particle at time t at position r, given that there is a particle at the origin at time 0). The results compared well with simulations of Brownian dynamics. Fraaije has performed a number of studies using DDFT to model the dynamics of block copolymer melts [160–162]. Dzubiella and Likos have used DDFT to study squeezing and relaxation of soft, Brownian particles in a time-dependent external field [163]. Their comparisons to Brownian dynamical simulations are very good. Van Teefelen, Likos, and Löwen used DDFT to study 2-D solid nucleation of particles interacting with an inverse cube potential [164]. Rex and Löwen have extended the theory for Brownian dynamics to include hydrodynamic interactions between the colloidal particles [165].

B. Energy Surface Methods and the Problem of Nucleation

One problem not easily treated with DDFT is that of systems crossing large energy barriers. The prototypical example of such a problem is the nucleation of one phase from another. For example, consider the problem of the nucleation of a vapor bubble in a superheated liquid. The superheated liquid is metastable, and random thermal fluctuations cause the formation of bubbles. These will either shrink and die if they are too small, or will grow without bound if they are large enough, thus converting the system from one phase to another. Such transitions are of interest in many important circumstances, and the question of nucleation has gained renewed interest due to the discovery of multistep processes involved in the nucleation of crystals from solution in the case of proteins [166]. However, even the physical picture just described for the nucleation of vapor bubbles has recently been called into question [167–169], thus contributing to the renewed interest in this subject. In the remainder of this subsection, a simple DFT for bubble nucleation will be presented and the classical nucleation theory will be reviewed. Various approaches to the description of nucleation within DFT will be described, concluding with a discussion of the current status of the subject.

As discussed previously, from any DFT —including exact DFT —a gradient expansion can be constructed. Let us parameterize the local density for the problem of bubble nucleation as a spherically symmetric, piecewise continuous function,

$$\rho(\mathbf{r}) = \bar\rho_0 \Theta(R-w-r)$$
$$+ \left(\bar\rho_0 + \frac{\bar\rho_\infty - \bar\rho_0}{2w}(r-R+w)\right)\Theta(R+w-r)\Theta(r-R+w) \quad (237)$$
$$+ \bar\rho_\infty \Theta(r-R)$$

which simply says that the density is ρ_0 for $r < R-w$ and ρ_∞ for $r > R+w$ and that it varies linearly in the intermediate region. The gradient model free energy will have the form

$$\Omega[\rho] = \int \left(f(\rho(\mathbf{r})) + g(\rho(\mathbf{r}))\left(\frac{\partial\rho(\mathbf{r})}{\partial\mathbf{r}}\right)^2 - \mu\rho(\mathbf{r})\right)d\mathbf{r} \quad (238)$$

where $f(\rho)$ is the (Helmholtz) free energy per unit volume in the bulk fluid. Substituting the parameterization into this expression gives a function, $F(\bar\rho_0, \bar\rho_\infty, R, w)$. However, this is still rather complicated so two simplifications are made. The first is to take the coefficient of the gradient to be independent of density, $g(\rho(\mathbf{r})) = g$. The second is to take the capillary approximation in which the width of the interface goes to zero, $w \to 0$, while keeping the combination $\gamma = g/(2w)^2$ constant. This then gives

$$\Omega = V(R)(f(\bar\rho_0)-\mu\bar\rho_0) + (V-V(R))(f(\bar\rho_\infty)-\mu\bar\rho_\infty) + \gamma(\bar\rho_\infty-\bar\rho_0)^2 S(R) \quad (239)$$

where $V(R) = \frac{4\pi}{3}R^3$, $S(R) = 4\pi R^2$ and V_0 is the overall volume of the system (assumed eventually to be infinite). The first two terms on the right are the free energy contributions of the bulk phases, and the third term is the contribution due to surface tension. For a given chemical potential, it is assumed that there is an equilibrium bulk vapor with density determined from $f'(\bar{\rho}_v) = \mu$ and an equilibrium liquid with density satisfying $f'(\bar{\rho}_l) = \mu$. [We will use the notation $\bar{\rho}_v(\mu)$ and $\bar{\rho}_l(\mu)$ to denote the solutions to these equations.] The Classical Nucleation Theory (CNT) uses this free energy function with $\bar{\rho}_0 = \bar{\rho}_v(\mu)$ and $\bar{\rho}_\infty = \bar{\rho}_l(\mu)$, so that the free energy is a function of a single parameter, R, the radius of the bubble. The function is cubic in the radius and has minima at zero radius (the metastable fluid), at some finite radius (the critical bubble) and at infinite radius (the uniform vapor). A slightly more general theory is to treat these densities as unknowns and to note that DFT tells us that the free energy function should be a minimized with respect to its parameters, giving

$$0 = (V-V(R))(f'(\bar{\rho}_\infty)-\mu) + 2\gamma(\bar{\rho}_\infty-\bar{\rho}_0)S(R)$$
$$0 = V(R)(f'(\bar{\rho}_0)-\bar{\rho}_0) - 2\gamma(\bar{\rho}_\infty-\bar{\rho}_0)S(R) \tag{240}$$
$$0 = R^2((f(\bar{\rho}_0)-\mu\bar{\rho}_0)-(f(\bar{\rho}_\infty)-\mu\bar{\rho}_\infty)) + 2\gamma(\bar{\rho}_\infty-\bar{\rho}_0)^2R$$

For a very large system, $V \gg V(R)$, the first equation implies that $0 = f'(\bar{\rho}_\infty)-\mu$, so that the outer density must be either $\bar{\rho}_v(\mu)$ or $\bar{\rho}_l(\mu)$: One chooses the former to describe bubble nucleation. The second and third equations then determine the values of the inner density and the radius. Note that one solution is $R = 0$, in which case the inner density is irrelevant and the system is bulk liquid. The solution for finite R corresponds to the critical radius. So long as $(f(\bar{\rho}_0)-\mu\bar{\rho}_0) < (f(\bar{\rho}_\infty)-\mu\bar{\rho}_\infty)$ and with V large, there will also be an extremum at $V(R) = V$ corresponding to the bulk vapor.

One aspect of nucleation that has long been treated with DFT is the determination of the height of the barrier to nucleation. This is because the barrier is defined by the critical nucleus — the metastable state the corresponds to a maximum of the free energy functional. The determination of this state from something like DFT goes back to Cahn and Hilliard [170], and calculations using more sophisticated DFT include the early work by Oxtoby and Evans [171] and Teng and Oxtoby [172]. (It is worth noting, however, that even in this case one is stretching DFT beyond its theoretical foundations. As shown in Section II, the only density to which unambiguous physical meaning can be attached is one that minimizes the free energy functional. Since the critical cluster *maximizes* the free energy functional, one is assuming, analogously to DDFT, that the free energy functional governs the dynamics of the transition and does not just define the equilibrium states.)

The question is how one might describe the transition from the liquid to the vapor—that is, all the points on the path that are not extrema of the free energy. In terms of the density, one notes that a system with a small bubble has more atoms, on average, than does a system with a large bubble since the gas density is lower than the liquid density. One might therefore try to stabilize a noncritical system by "minimizing the free energy for a fixed number of atoms." This means minimizing the free energy subject to the constraint

$$\int \rho(\mathbf{r})d\mathbf{r} = N \tag{241}$$

for some specified value of N. The natural way to do this is through the method of Lagrange multipliers. One forms the Lagrangian,

$$L(\alpha, [\rho]) = \Omega[\rho] - \alpha \left(\int \rho(\mathbf{r})d\mathbf{r} - N \right) \tag{242}$$

and then solves for the constrained minimum via

$$\frac{\delta L(\alpha, [\rho])}{\delta \rho(\mathbf{r})} = 0, \qquad \frac{\partial L(\alpha, [\rho])}{\partial \alpha} = 0 \tag{243}$$

giving

$$\frac{\delta F[\rho]}{\delta \rho(\mathbf{r})} - \mu - \alpha = 0$$

$$\int \rho(\mathbf{r}; \mu, \alpha)d\mathbf{r} = N \tag{244}$$

Thus, the effect of this procedure is first to shift the chemical potential from μ to $\mu + \alpha$, then to solve for the extremum for the shifted chemical potential, and then to adjust α until the system has the desired number of atoms. The result of this calculation is simply the critical cluster for the shifted chemical potential $\mu' = \mu + \alpha$. Note in particular that for the example of bubble nucleation, the density far from the bubble will not be $\bar{\rho}_l(\mu)$ but, rather, $\bar{\rho}_l(\mu')$. Away from the critical point, the vapor density is quite small, so that the difference between $\bar{\rho}_v(\mu)$ and $\bar{\rho}_v(\mu + \alpha)$ is quite small. Thus, for the problem of *liquid droplet* nucleation in a vapor, this shift of the background is probably not too important. However, for the problem of *bubble nucleation* in a fluid, the shift from $\bar{\rho}_l(\mu)$ to $\bar{\rho}_l(\mu + \alpha)$ can be quite substantial, potentially affecting the physics. For this reason, other methods have been developed.

One variation of the constraint approach is to fix the number of atoms within some volume V_0 which is smaller than the system volume V,

$$\int_{V_0} \rho(\mathbf{r})d\mathbf{r} = N \tag{245}$$

In this case, there are two parameters, the radius of the constrained volume, R_0, and the number N. The Euler–Lagrange equation now takes the form

$$\frac{\delta F[\rho]}{\delta \rho(\mathbf{r})} - \mu - \alpha \Theta(R_0 - r) = 0 \tag{246}$$

so that in effect one has a discontinuous chemical potential. This necessarily leads
to discontinuous density profiles unless $\alpha = 0$, which only happens for the critical
cluster for the chemical potential μ: All subcritical and supercritical clusters will
exhibit discontinuous profiles. Talanquer and Oxtoby [173] used this method to
study droplet nucleation. However, rather than solve the Euler–Lagrange equation,
Eq. (246), for $r > R_0$, they took V_0 to be much larger than the droplet and
approximated the outer density as a constant that was adjusted so as to give a
continuous density profile. This would appear to essentially be the same as the
calculation described above with a shifted chemical potential everywhere. Uline
and Corti used Eq. (246) as it stands, thereby obtaining discontinuous profiles [167]. Their results are particularly interesting because they report that
supercritical bubbles become unstable so that no solution to the equations exists
above some critical size. Their conclusion is that the assumption that the outer
density is $\bar{\rho}_l(\mu)$ is no longer true, thus indicating something like a spinodal
breakdown of the fluid. However, it is hard to see, physically, how a bubble of finite
size can destabilize the fluid far away from the bubble. In fact, it was subsequently
shown that using this constraint with the toy DFT given above, similar instabilities
occur [168, 169]. This is disturbing because Eq. (239) is a smooth, continuous
function of its arguments, so any such instability must be due to the constraint.
Furthermore, another physically reasonable constraint was shown not to give such
instabilities, so that the robustness of the constraint method must be questioned.

    An alternative method has recently been proposed which appears to circumvent
such problems [168, 169]. Just as classical DFT was inspired by methods
developed in the context of ab initio quantum mechanical calculations, so the
new technique is borrowed from the quantum chemistry community. In quantum
chemistry, a subject of great interest is the determination of reaction pathways—
that is, a description of how a complicated, many-body system transforms
from one metastable configuration into another. What is known is the free energy
surface governing the transition. Many methods have been developed, primarily
within the last 20 years, for finding the most probable path for such transitions
(see, e.g., Wales [174] and references therein). The problem of nucleation in
classical systems is quite similar in that it is formulated as the desire to trace the
path in density space by which the system transforms from a metastable uniform
state to another, stable, uniform state. If one accepts the functional $\Omega[\rho]$ as
governing the dynamics of the transition—as is done, for example, in DDFT—
then the same energy surface methods can be used.

    One approach is to consider a collection of $M + 1$ density functions,
$\{\rho^{(m)}(\mathbf{r})\}_{m=0}^{M}$, where $\rho^{(0)}(\mathbf{r})$ is the initial metastable state, $\rho^{(M)}(\mathbf{r})$ is the final
stable state, and the remaining states define a path between these two. For example,
in the bubble nucleation problem, one has that $\rho^{(0)}(\mathbf{r}) = \bar{\rho}_l(\mu)$ and
$\rho^{(M)}(\mathbf{r}) = \bar{\rho}_v(\mu)$. In between, in the toy model, one can choose the parameters

for the mth density, $\bar{\rho}_0^{(m)}$ and $R^{(m)}$, in a convenient way, say as $\bar{\rho}_0^{(m)} = \bar{\rho}_v(\mu)$ and $R^{(m)} = \frac{m}{M} R_V$ where R_V is the radius of the overall volume V. The goal is to adjust this chain of images, as they are called, so as to map out the most likely path between the endpoints in density space. As discussed in Wales [174], there are several ways to do this and here, a particularly simple one called the Nudged Elastic Band (NEB) [175] method is described.

Clearly, if one tried to minimize the total energy of the path, $E = \sum_{m=1}^{M-1} \Omega[\rho^{(m)}]$, the densities would all end up in one of the metastable states because these are, by definition, local minima. To map out the desired path, it is is necessary to force the densities to remain evenly spaced, in some sense, along the path. In the NEB, this is done by adding fictitious elastic forces between neighboring images. However, simply adding the couplings is too crude because the elastic forces alter the effective free energy landscape. Instead, one wants to apply the elastic forces only *along* the path and to minimize the total energy in directions perpendicular to the path, with no elastic forces. The key then is to define, for each image, the direction in density space of the tangent to the current path. First, one defines an inner product and distance in density space,

$$\langle \rho^{(m)}, \rho^{(m')} \rangle = \int \rho^{(m)}(\mathbf{r}) \rho^{(m')}(\mathbf{r}) d\mathbf{r}$$

$$d[\rho^{(m)}, \rho^{(m')}] = \langle \rho^{(m)} - \rho^{(m')}, \rho^{(m)} - \rho^{(m')} \rangle^{1/2} \tag{247}$$

$$= \left(\int (\rho^{(m)}(\mathbf{r}) - \rho^{(m')}(\mathbf{r}))^2 d\mathbf{r} \right)^{1/2}$$

The tangent at image m is defined in terms of its neighbors, images $m-1$ and $m+1$, based on the local energy landscape. For example, if the energy is monotonically increasing, $\Omega[\rho^{(m-1)}] < \Omega[\rho^{(m)}] < \Omega[\rho^{(m+1)}]$, then the tangent at the image ρ_m, called t_m, is

$$t^{(m)}(\mathbf{r}) = \rho^{(m+1)}(\mathbf{r}) - \rho^{(m)}(\mathbf{r}) \tag{248}$$

and the normalized tangent, $\hat{t}^{(m)}(\mathbf{r}) = t^{(m)}(\mathbf{r}) / \langle t^{(m)}, t^{(m)} \rangle$. If the energy is monotonically decreasing, the tangent is based on $\rho^{(m)} - \rho^{(m-1)}$. For nonmonotonic neighbors, the heuristic is given in Ref. 175. The NEB method then consists of finding a configuration that gives zero NEB force. Let the "force" due to the actual free-energy surface be $\mathcal{F}^{(m)}(\mathbf{r}) = -\frac{\partial \beta \Omega[\rho^{m(i)}]}{\partial \rho(\mathbf{r})}$. Then the NEB method consists of solving

$$0 = \mathcal{F}^{\perp(m)}(\mathbf{r}) + k\hat{t}^{(m)}(\mathbf{r})(d[\rho^{(m+1)}, \rho^{(m)}] - d[\rho^{(m)}, \rho^{(m-1)}]) \tag{249}$$

where $\mathcal{F}^{\perp(m)}(\mathbf{r}) = \mathcal{F}^{(m)}(\mathbf{r}) - \hat{t}^{(m)}(\mathbf{r})(\hat{t}^{(m)} * \mathcal{F}^{(m)})$ is the component of the thermodynamic force orthogonal to the tangent vector and k is the spring constant. Further details can be found in Refs. 168 and 169, where the NEB method is applied to the problems of bubble and droplet nucleation. For droplet nucleation, where a

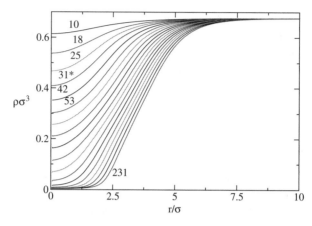

Figure 9. The process of bubble formation as calculated using the NEB and DFT for a Lennard-Jones fluid with $k_B T = 0.8\varepsilon$ and supersaturation $\frac{\mu - \mu_{coex}}{\mu_{coex}} = -0.20$, where μ_{coex} is the value of the chemical potential at coexistence. The potential is truncated and shifted at a distance $r_c = 4\sigma$. The figure shows the density profile, $\rho(r)$, for the various images along the optimal path between the uniform liquid and the uniform gas. The size, in terms of the number of missing atoms relative to the background liquid, is given for several profiles and the critical profile is marked with an asterisk. From Lutsko [169].

significant amount of detailed simulation data is available, the method proves to be very accurate in locating and describing the critical cluster as well as in computing the shape of nucleation barrier. No sign of instabilities is found in these calculations. As shown in Fig. 9, the picture of bubble formation that emerges from these calculations is very different than that of CNT. Rather than forming as a gas-containing void with initially small radius that slowly grows, the bubble forms as a finite-sized region in which the density gradually decreases until it reaches something approaching the gas density. Only then does it begin to grow in radius; in the example shown, the growth occurs only long after the bubble has passed criticality. Very recently, a similar technique (but based on the String method [176], rather than the NEB) has been applied to the problem of capillary condensation [177].

VII. CONCLUSIONS

The main points to be drawn from the survey of the current state of DFT are as follows:

1. Classical DFT is based on a collection of exact theorems regarding the behavior of systems in the grand canonical ensemble. In some cases, such as the ideal gas, the quasi-zero-dimensional system, and one-dimensional hard rods, the exact excess free energy functional can be constructed.

2. Effective liquid theories and liquid-state-based perturbation theory continue to be used in many calculations because of their simplicity.

3. For realistic, three-dimensional, inhomogeneous systems, the best theory currently available is probably the Fundamental Measure Theory for hard spheres.

4. Most calculations for potentials other than hard spheres are performed using a sum of the hard-sphere excess free energy and a perturbative and/or mean-field treatment of the attractive part of the potential.

5. DFT is increasingly used to study non-equilibrium systems. Many models dating back to the 1970s include something like a local pressure expressed as a functional derivative of the free energy. More recently, Dynamic DFT, a generalization of the diffusion equation, is used to model overdamped systems. Other methods for exploring the energy surface are also being used to study problems involving large barriers such as homogeneous nucleation.

Acknowledgments

This work was supported in part by the European Space Agency under contract number ESA AO-2004-070.

APPENDIX A: DFT FOR HARD RODS

The statistical mechanics of hard rods can be completely solved as discussed, for example, in Ref. 70. Surprisingly, it was only in 1976 that Percus first explicitly worked out the structure of DFT for hard rods [9]. The presentation here follows his work closely.

The great simplification of hard spheres in one dimension is that (a) they only interact with there two nearest neighbors and (b) they cannot move past one another. The second property means that we can, without loss of generality, label a finite collection of hard rods such that $q_1 < q_2 < \ldots < q_n$. Hence, if the system is subject to a one-body potential, $\phi(r)$, the grand partition function is

$$
\Xi(\beta, \mu; [\phi]) = 1 + \int_{-\infty}^{\infty} \exp(-\beta\tilde{\phi}(q_1))dq_1
$$
$$
+ \sum_{n=2}^{\infty} \int_{-\infty}^{\infty} \exp\left(-\beta\sum_{i=1}^{n}\tilde{\phi}(q_i)\right) W(q_1, \ldots, q_n)dq_1, \ldots, dq_n
\tag{A1}
$$

where $\tilde{\phi}(r) \equiv \phi(r) - \mu$ and

$$
W(q_1, \ldots, q_n) \equiv \Theta(q_2 - q_1 - d) \ldots \Theta(q_n - q_{n-1} - d)
\tag{A2}
$$

The one-body density is

$$\rho(r) = \langle \hat{\rho}(r) \rangle = -k_B T \frac{\delta \ln \Xi}{\delta \phi(r)} \tag{A3}$$

$$= \Xi^{-1} \exp(-\beta \tilde{\phi}(r))$$

$$+ \Xi^{-1} \sum_{n=2}^{\infty} \sum_{k=1}^{n} \left\{ \int_{-\infty}^{\infty} \exp\left(-\beta \sum_{i=1}^{k-1} \tilde{\phi}(q_i) \right) \right.$$

$$\times W(q_1, \dots, q_{k-1}, r) dq_1, \dots, dq_{k-1} \}^{(k=1)} \tag{A4}$$

$$\times \exp(-\beta \tilde{\phi}(r)) \left\{ \int_{-\infty}^{\infty} \exp\left(-\beta \sum_{i=k+1}^{n} \tilde{\phi}(q_i) \right) \right.$$

$$\times W(r, q_{k+1}, \dots, q_n) dq_{k+1}, \dots, dq_n \}^{(k=n)} \tag{A5}$$

where the notation $\{\dots\}^{(k=r)}$ means that the bracket should be set equal to 1 if $k = r$. Relabeling the dummy integration variables $q_{k+1}, \dots, q_n$ allows the last term in brackets to be written as

$$\int_{-\infty}^{\infty} \exp\left(-\beta \sum_{i=k+1}^{n} \tilde{\phi}(q_i) \right) W(r, q_{k+1}, \dots, q_n) dq_{k+1}, \dots, dq_n$$

$$= \int_{-\infty}^{\infty} \exp\left(-\beta \sum_{i=1}^{n-k} \tilde{\phi}(q_i) \right) W(r, q_1, \dots, q_{n-k}) dq_1 \dots dq_{n-k} \tag{A6}$$

Then, noting that

$$\sum_{n=1}^{\infty} \sum_{k=1}^{n} X(k,n) = \sum_{k=1}^{\infty} \sum_{n=k}^{\infty} X(k,n) = \sum_{k=1}^{\infty} \sum_{m=1}^{\infty} X(k,m+k-1) \tag{A7}$$

allows this to be written as

$$\rho(r) = \Xi^{-1} \sum_{k=1}^{\infty} \sum_{m=1}^{\infty} \exp(-\beta\mu(m+k-1))$$

$$\times \left\{ \int_{-\infty}^{\infty} \exp\left(-\beta \sum_{i=1}^{k-1} \tilde{\phi}(q_i) \right) W(q_1, \dots, q_{k-1}, r) dq_1 \dots dq_{k-1} \right\}^{(k=1)}$$

$$\times \exp(-\beta\phi(r)) \left\{ \int_{-\infty}^{\infty} \exp\left(-\beta \sum_{i=1}^{m-1} \tilde{\phi}(q_i) \right) \right.$$

$$\times W(r, q_1, \dots, q_{m-1}) dq_1 \dots dq_{m-1} \}^{(m=1)} \tag{A8}$$

which clearly factorizes into the product of terms that resemble the grand partition function. It is convenient to define a more general function of two variables as

$$\Xi(x,y;\beta,\mu;[\phi])$$

$$= \Theta(y-(x+d))\sum_{n=1}^{\infty}\left\{\int_{-\infty}^{\infty}\exp\left(-\beta\sum_{i=1}^{n-1}\tilde{\phi}(q_i)\right)\right.$$

$$\left.\times W(x,q_1,\ldots,q_{n-1},y)dq_1,\ldots,dq_{n-1}\right\}^{(n=1)}$$

$$= \Theta(y-(x+d))+\Theta(y-(x+d))\sum_{k=1}^{\infty}\int_{-\infty}^{\infty}\exp\left(-\beta\sum_{i=1}^{k}\tilde{\phi}(q_i)\right)$$

$$\times W(x,q_1,\ldots,q_k,y)dq_1\ldots dq_k \tag{A9}$$

so that $\Xi(\beta,\mu;[\phi]) = \Xi(-\infty,\infty;\beta,\mu;[\phi])$. Then, one has that

$$\rho(r) = \exp\left(-\beta\tilde{\phi}(r)\right)\Xi(-\infty,r;\beta,\mu;[\phi])\Xi(r,\infty;\beta,\mu;[\phi])\Xi^{-1}(-\infty,\infty;\beta,\mu;[\phi]) \tag{A10}$$

Now, simple differentiation gives

$$\frac{\partial}{\partial x}\Xi(x,y) = -\delta(y-(x+d))-\exp\left(-\beta\tilde{\phi}(x+d)\right)\Xi(x+d,y)$$

$$\frac{\partial}{\partial y}\Xi(x,y) = \delta(y-(x+d))+\exp\left(-\beta\tilde{\phi}(y-d)\right)\Xi(x,y-d) \tag{A11}$$

and, in particular,

$$\frac{\partial}{\partial r}\Xi(r,\infty) = -\exp\left(-\beta\tilde{\phi}(r+d)\right)\Xi(r+d,\infty)$$

$$= -\rho(r+d)\Xi(-\infty,\infty)/\Xi(-\infty,r+d)$$

$$\frac{\partial}{\partial r}\Xi(-\infty,r)= \exp\left(-\beta\tilde{\phi}(r-d)\right)\Xi(-\infty,r-d)$$

$$= \rho(r-d)\Xi(-\infty,\infty)/\Xi(r-d,\infty) \tag{A12}$$

Note that the second equation implies

$$\frac{\partial}{\partial r}\Xi(-\infty,r+d) = \rho(r)\Xi(-\infty,\infty)/\Xi(r,\infty) \tag{A13}$$

Combined with the first line of Eq. (A13), one has that

$$\frac{\partial}{\partial r}\Xi(r,\infty)\Xi(-\infty,r+d) = (\rho(r)-\rho(r+d))\Xi(-\infty,\infty) \tag{A14}$$

so

$$\Xi(x,\infty)\Xi(-\infty,x+d) = \Xi(-\infty,\infty) - \Xi(-\infty,\infty)\int_x^{x+d}\rho(r)dr \qquad (A15)$$

where the integration constant is fixed by assuming that the density vanishes sufficiently fast as $r \to \infty$, the limit $x \to \infty$. With this result, Eq. (A13) can now be written as

$$\frac{\partial}{\partial r}\Xi(r,\infty) = \frac{-\rho(r+d)\Xi(r,\infty)}{1-\int_r^{r+d}\rho(y)dy}$$

$$\frac{\partial}{\partial r}\Xi(-\infty,r) = \frac{\rho(r-d)\Xi(-\infty,r)}{1-\int_{r-d}^r\rho(y)dr} \qquad (A16)$$

giving

$$\ln\Xi(x,\infty) = \ln\Xi(-\infty,\infty) - \int_{-\infty}^x \frac{\rho(r+d)}{1-\int_r^{r+d}\rho(y)dy}dr$$

$$\ln\Xi(-\infty,x) = \ln\Xi(-\infty,\infty) - \int_x^\infty \frac{\rho(r-d)}{1-\int_{r-d}^r\rho(y)dy}dr \qquad (A17)$$

where, again, the integration constants are fixed by taking the limit $x \to \mp\infty$. The opposite limits, $x \to \pm\infty$, then give

$$\ln\Xi(-\infty,\infty) = \int_{-\infty}^\infty \frac{\rho(r+d)}{1-\int_r^{r+d}\rho(y)dy}dr = \int_{-\infty}^\infty \frac{\rho(r+d/2)}{1-\int_{-d/2}^{d/2}\rho(r+y)dy}dr$$

$$\ln\Xi(-\infty,\infty) = \int_{-\infty}^\infty \frac{\rho(r-d)}{1-\int_{r-d}^r\rho(y)dy}dr = \int_{-\infty}^\infty \frac{\rho(r-d/2)}{1-\int_{-d/2}^{d/2}\rho(r+y)dy}dr \qquad (A18)$$

These can be combined to give the symmetric form:

$$\ln\Xi(-\infty,\infty) = \int_{-\infty}^\infty \frac{\frac{1}{2}(\rho(r+d/2)+\rho(r-d/2))}{1-\int_{-d/2}^{d/2}\rho(r+y)dy}dr \qquad (A19)$$

Extensions to sticky hard spheres and to mixtures of hard spheres have also been given [71, 178].

To make contact with the density functional formalism, note that the field does not occur in this expression: It is the equivalent of Eq. (23) giving the grand potential at equilibrium, after the field has been eliminated. To get $F[n]$, we use

Eq. (A15) in Eq. (A10) and take the log to get

$$
\frac{\delta \beta F[\rho]}{\delta \rho(\mathbf{r})} = -\beta \tilde{\phi}(r) = \ln \rho(r) - \ln \Xi(-\infty, r) - \ln \Xi(r, \infty) + \ln \Xi(-\infty, \infty)
$$

$$
= \ln \rho(r) + \int_{-\infty}^{r} \left(-\frac{\rho(x-d)}{1 - \int_{x-d}^{x} \rho(y) dy} + \frac{\rho(x+d)}{1 - \int_{x}^{x+d} \rho(y) dy} \right) dr
$$

$$
= \ln \rho(r) - \int_{-\infty}^{r-d/2} \left(\frac{\rho(x-d/2)}{1 - \int_{-d/2}^{d/2} \rho(x+y) dy} \right) dr
$$

$$
+ \int_{-\infty}^{r+d/2} \left(\frac{\rho(x+d/2)}{1 - \int_{-d/2}^{d/2} \rho(x+y) dy} \right) dr
$$

$$
= \ln \rho(r) - \frac{1}{2} \ln \left(1 - \int_{r-d}^{r} \rho(y) dy \right) dr - \frac{1}{2} \ln \left(1 - \int_{r}^{r+d} \rho(y) dy \right)
$$

$$
+ \frac{1}{2} \int_{-r-d/2}^{r+d/2} \left(\frac{\rho(x+d/2) + \rho(x-d/2)}{1 - \int_{-d/2}^{d/2} \rho(x+y) dy} \right) dr \qquad \text{(A20)}
$$

Performing the functional integration gives

$$
\beta F[\rho] = \beta F_{\text{id}}[\rho] - \int \frac{1}{2} (\rho(x+d/2) + \rho(x-d/2)) \ln \left(1 - \int_{-d/2}^{d/2} \rho(x+y) dy \right) dr
$$
$$
\text{(A21)}
$$

In principle, there could be a density-independent integration constant, but this is of no relevance.

APPENDIX B: FMT TWO-BODY TERM

The FMT two-body contribution is

$$
F_2 = - \int d\mathbf{r} \, \psi(\eta(\mathbf{r})) \int ds_1 ds_2 \, \rho(\mathbf{r} + \mathbf{s}_1) w(s_1) \rho(\mathbf{r} + \mathbf{s}_2) w(s_2) P(\mathbf{s}_1, \mathbf{s}_2) \quad \text{(B1)}
$$

The effect of the kernel $P(\mathbf{s}_1, \mathbf{s}_2)$ is to couple what would otherwise be two factors of $s(\mathbf{r})$. To see what kind of contribution this gives when the density is a sum of delta functions, consider the integral

$$
\int ds_1 ds_2 \, \delta(\mathbf{r} + \mathbf{s}_1) w(s_1) \delta(\mathbf{r} + \mathbf{s}_2 - \mathbf{R}) w(s_2) P(\mathbf{s}_1, \mathbf{s}_2)
$$

$$
= \left(\frac{1}{4\pi(d/2)^2} \right)^2 \delta \left(r - \frac{d}{2} \right) \delta \left(|\mathbf{R} - \mathbf{r}| - \frac{d}{2} \right) P(-\mathbf{r}, \mathbf{R} - \mathbf{r}) \qquad \text{(B2)}
$$

Since the kernel is a scalar, assume it is a function of the scalar product of its arguments, $P(\mathbf{s}_1, \mathbf{s}_2) = P(\mathbf{s}_1 \cdot \mathbf{s}_2)$. Let $\mathbf{R}$ define the z-direction in the r integral so that $\mathbf{R} \cdot \mathbf{r} = Rrx$ where x is the azimuthal variable. Using the fact that the first delta function fixes the value of r and using the usual rules for change of variable in the second delta function gives

$$
\int d\mathbf{s}_1 d\mathbf{s}_2 \, \delta(\mathbf{r}+\mathbf{s}_1) w(s_1) \delta(\mathbf{r}+\mathbf{s}_2-\mathbf{R}) w(s_2) P(\mathbf{s}_1, \mathbf{s}_2)
$$
$$
= \left(\frac{1}{4\pi(d/2)^2}\right)^2 \delta\left(r-\frac{d}{2}\right) \frac{\delta(x-R/d)}{R} P\left(\left(\frac{d}{2}\right)^2 - \left(\frac{R^2}{2}\right)\right) \tag{B3}
$$

Assuming that $\lim_{R \to 0} P\left(\left(\frac{d}{2}\right)^2 - \left(\frac{R^2}{2}\right)\right)/R = 0$, one immediately finds

$$
F_2 = -2N_1 N_2 \left(\frac{1}{4\pi(d/2)^2}\right)^2 R^{-1} P\left(\left(\frac{d}{2}\right)^2 - \left(\frac{R^2}{2}\right)\right)
$$
$$
\times \int d\mathbf{r} \, \psi(\eta(\mathbf{r})) \delta\left(r-\frac{d}{2}\right) \delta(x-R/d)
$$
$$
= -2N_1 N_2 \left(\frac{1}{4\pi(d/2)^2}\right)^2 P\left(\left(\frac{d}{2}\right)^2 - \left(\frac{R^2}{2}\right)\right) \tag{B4}
$$
$$
\times \int d\mathbf{r} \, \psi(\eta(\mathbf{r})) \delta\left(r-\frac{d}{2}\right) \delta\left(|\mathbf{R}-\mathbf{r}|-\frac{d}{2}\right)
$$

This integral is easily evaluated as a limit of the case that the cavities have different diameters, d_1 and d_2. Writing $\psi(\eta) = \frac{\partial^2}{\partial \eta^2} \psi_0(\eta)$ gives

$$
F_2 = -2\left(\frac{1}{4\pi(d/2)^2}\right)^2 P\left(\left(\frac{d}{2}\right)^2 - \left(\frac{R^2}{2}\right)\right)
$$
$$
\times \lim_{d_1, d_2 \to d} \frac{\partial^2}{\partial(d_1/2)\partial(d_2/2)} \int d\mathbf{r} \, \psi_0(\eta(\mathbf{r}))
$$
$$
= -2\left(\frac{1}{4\pi(d/2)^2}\right)^2 P\left(\left(\frac{d}{2}\right)^2 - \left(\frac{R^2}{2}\right)\right) \tag{B5}
$$
$$
\times \lim_{d_1, d_2 \to d} 4 \frac{\partial^2}{\partial d_1 \partial d_2} (|V_1 - V_{12}|\psi_0(N_1) + |V_2 - V_{12}|\psi_0(N_2) + |V_{12}|\psi_0(N))
$$
$$
= -2\left(\frac{1}{4\pi(d/2)^2}\right)^2 P\left(\left(\frac{d}{2}\right)^2 - \left(\frac{R^2}{2}\right)\right) \frac{\pi d^2}{2R} (\psi_0(N) - \psi_0(N_1) - \psi_0(N_2))
$$

The sum gives the exact result, provided that $\psi_0(\eta) = \Phi_0(\eta)$ and

$$-2\left(\frac{1}{4\pi(d/2)^2}\right)^2 P\left(\left(\frac{d}{2}\right)^2 - \frac{R^2}{2}\right)\frac{\pi d^2}{2R} = \frac{R}{d} \tag{B6}$$

Simplification gives

$$P\left(\frac{d^2}{4} - \frac{R^2}{2}\right) = -\pi d R^2 \tag{B7}$$

or

$$P(y) = -\pi d\left(\frac{d^2}{2} - 2y\right) = -4\pi \frac{d}{2}\left(\frac{d^2}{4} - y\right) \tag{B8}$$

Putting this together gives the two-body contribution

$$\begin{aligned}
F_2 &= 4\pi \frac{d}{2}\left(\frac{1}{4\pi(d/2)^2}\right)^2 \int d\mathbf{r}\,\psi''_0(\eta(\mathbf{r})) \\
&\quad \times \int d\mathbf{s}_1 d\mathbf{s}_2\, \rho(\mathbf{r}+\mathbf{s}_1)w(\mathbf{s}_1)\rho(\mathbf{r}+\mathbf{s}_2)w(\mathbf{s}_2)\left(\frac{d^2}{4} - \mathbf{s}_1 \cdot \mathbf{s}_2\right) \\
&= \frac{1}{2\pi d}\int d\mathbf{r}\,\psi''_0(\eta(\mathbf{r}))\left(s^2(\mathbf{r}) - v^2(\mathbf{r})\right)
\end{aligned} \tag{B9}$$

which is the same as the second contribution to the Rosenfeld functional.

APPENDIX C: PROOF OF THE WALL THEOREM FOR THE VDW MODEL

Here, the proof of the wall theorem is given for VdW models such as given in Eqs. (190)–(193). The location of the wall is taken to be $z = 0$, and the system is uniform in the other directions. It therefore follows that the potential and the FMT weights can be integrated over the directions parallel to the wall and all quantities become one-dimensional functions of z [see, e.g., Eq. (163)]. Restricting attention to the region $z > 0$, the Euler–Lagrange equation is

$$0 = \ln\rho(z_1) + \int dz_2 \frac{\partial\Phi(z_2;[\rho])}{\partial n_\alpha(z_2)} w_\alpha(z_{21}) - \mu + \int_{-\infty}^{\infty} w(z_{12})\rho(z_2)dz_2 \tag{C1}$$

Differentiating with respect to z_1 and multiplying through by $\rho(z_1)$ gives

$$
0 = \frac{d\rho(z_1)}{dz_1} + \int_{-\infty}^{\infty} dz_2 \frac{\partial\Phi(z_2;[\rho])}{\partial n_\alpha(z_2)} \rho(z_1) \frac{d}{dz_1} w_\alpha(z_{21})
$$
$$
+ \int_{-\infty}^{\infty} \rho(z_1)\left(\frac{d}{dz_1} w(z_{12})\right)\rho(z_2)dz_2 \tag{C2}
$$

Now, we make two important assumptions. First is that there is some point $z_B > 0$ such that for $z > z_B$, the density is indistinguishable from the bulk: $\rho(z) = \rho_l(\mu)$, where the bulk density is determined by the imposed chemical potential. Second, we assume that the potential tail has a finite range r_c. We now integrate Eq. (C2) from an initial point at the wall, 0_+, to a point z_b that is sufficiently far in the bulk region that $z_b > z_B + z_c$ and $z_b > z_B + d$. This gives

$$
0 = \rho(z_b) - \rho(0_+) + \int_{-\infty}^{\infty} dz_2 \frac{\partial\Phi(z_2;[\rho])}{\partial n_\alpha(z_2)} \int_{0_+}^{z_b} \rho(z_1) \frac{d}{dz_1} w_\alpha(z_{21}) dz_1
$$
$$
+ \int_{0_+}^{z_b} dz_1 \int_0^{\infty} \rho(z_1)\left(\frac{d}{dz_1} w(z_{12})\right)\rho(z_2)dz_2 \tag{C3}
$$

where the fact that $\rho(z) = 0$ for $z < 0$ has been used. Now, by assumption, $\rho(z_b) = \rho_l(\mu)$, so this can be rearranged to give

$$
\rho(0_+) = \rho_l(\mu) + I_{\text{FMT}} + I_{\text{MF}} \tag{C4}
$$

The mean field contribution is

$$
I_{\text{MF}} = \int_{0_+}^{z_b} dz_1 \int_{0_+}^{\infty} \rho(z_1)\left(\frac{d}{dz_1} w(z_{12})\right)\rho(z_2)dz_2
$$
$$
= \int_{0_+}^{z_b} dz_1 \int_{0_+}^{z_b} \rho(z_1)\left(\frac{d}{dz_1} w(z_{12})\right)\rho(z_2)dz_2 \tag{C5}
$$
$$
+ \int_{0_+}^{z_b} dz_1 \int_{z_b}^{\infty} \rho(z_1)\left(\frac{d}{dz_1} w(z_{12})\right)\rho(z_2)dz_2
$$

The first term on the right vanishes since it is odd under a relabeling $z_1 \leftrightarrow z_2$. Thus

$$
I_{\text{MF}} = \int_{0_+}^{z_b} dz_1 \int_{z_b}^{\infty} \rho(z_1)\left(\frac{d}{dz_1} w(z_{12})\right)\rho(z_2)dz_2
$$
$$
= \rho_l(\mu) \int_{0_+}^{z_b} dz_1 \int_{z_b}^{\infty} \rho(z_1)\left(\frac{d}{dz_1} w(z_{12})\right)dz_2
$$
$$
= \rho_l(\mu) \int_{0_+}^{z_b} \rho(z_1) w(z_1 - z_b)dz_1 \tag{C6}
$$

Making use of the finite range of the potential

$$I_{MF} = \rho_l(\mu) \int_{z_b-r_c}^{z_b} \rho(z_1) w(z_1-z_b) dz_1 \tag{C7}$$

and since, by hypothesis, z_b-r_c is still in the bulk region, this gives

$$
\begin{aligned}
I_{MF} &= \rho_l(\mu)^2 \int_{z_b-r_c}^{z_b} w(z_1-z_b) dz_1 \\
&= \rho_l(\mu)^2 \int_0^{r_c} w(z_1) dz_1
\end{aligned}
\tag{C8}
$$

The FMT contribution is

$$
\begin{aligned}
I_{FMT} &= \int_{-\infty}^{\infty} dz_2 \frac{\partial \Phi(z_2;[\rho])}{\partial n_\alpha(z_2)} \int_{0_+}^{z_b} \rho(z_1) \frac{d}{dz_1} w_\alpha(z_{21}) dz_1 \\
&= -\int_{-\infty}^{\infty} dz_2 \frac{\partial \Phi(z_2;[\rho])}{\partial n_\alpha(z_2)} \int_{0_+}^{\infty} \rho(z_1) \frac{d}{dz_2} w_\alpha(z_{21}) dz_1 \\
&\quad + \int_{-\infty}^{\infty} dz_2 \frac{\partial \Phi(z_2;[\rho])}{\partial n_\alpha(z_2)} \int_{z_b}^{\infty} \rho(z_1) \frac{d}{dz_2} w_\alpha(z_{21}) dz_1 \\
&= -\int_{-\infty}^{\infty} dz_2 \frac{\partial \Phi(z_2;[\rho])}{\partial n_\alpha(z_2)} \frac{dn_\alpha(z_2)}{dz_2} \\
&\quad + \rho_l(\mu) \int_{-\infty}^{\infty} dz_2 \frac{\partial \Phi(z_2;[\rho])}{\partial n_\alpha(z_2)} \int_{z_b}^{\infty} \frac{d}{dz_2} w_\alpha(z_{21}) dz_1
\end{aligned}
\tag{C9}
$$

The first term is an exact differential

$$\int_{-\infty}^{\infty} dz_2 \frac{\partial \Phi(z_2;[\rho])}{\partial n_\alpha(z_2)} \int_{0_+}^{\infty} \rho(z_1) \frac{d}{dz_2} w_\alpha(z_{21}) dz_1 = f_{ex}^{HS}(\rho_\infty d^3) - f_{ex}^{HS}(\rho_{-\infty} d^3) \tag{C10}$$

where $f_{ex}^{HS}(\rho d^3)$ is the excess free energy per unit volume in the hard-sphere fluid. The second term is

$$
\begin{aligned}
\rho_l(\mu) \int_{-\infty}^{\infty} dz_2 \frac{\partial \Phi(z_2;[\rho])}{\partial n_\alpha(z_2)} \int_{z_b}^{\infty} \frac{d}{dz_2} w_\alpha(z_{21}) dz_1 \\
= \rho_l(\mu) \int_{-\infty}^{\infty} dz_2 \frac{\partial \Phi(z_2;[\rho])}{\partial n_\alpha(z_2)} w_\alpha(z_2-z_b)
\end{aligned}
\tag{C11}
$$

Now, the function $w_\alpha(z_{21})$ is only nonzero in the range $-d/2 < z_{21} < d/2$; so as long as $z_b - d/2$ is in the bulk region, one has that

$$\rho_l(\mu) \int_{-\infty}^{\infty} dz_2 \frac{\partial \Phi(z_2; [\rho])}{\partial n_\alpha(z_2)} \int_{z_b}^{\infty} \frac{d}{dz_2} w_\alpha(z_{21}) dz_1 = \rho_l(\mu) \frac{\partial f_{HS}(\rho_b d^3)}{\partial \rho_b} \quad \text{(C12)}$$

Thus

$$I_{FMT} = -f_{ex}^{HS}(\rho_l(\mu)d^3) + \rho_l(\mu) \frac{\partial f_{ex}^{HS}(\rho_l(\mu)d^3)}{\partial \rho_l(\mu)} = \beta P_{ex}^{HS}(\rho_l(\mu)d^3) \quad \text{(C13)}$$

The final result is

$$\rho(0_+) = \rho_l(\mu) + \beta P_{ex}^{HS}(\rho_l(\mu)d^3) + \rho_l(\mu)^2 \int_0^{r_c} w(z_1) dz_1 = \beta P(\rho_l(\mu)) \quad \text{(C14)}$$

References

1. J. S. Rowlinson, *J. Stat. Phys.* **20**, 197 (1979).

2. J. D. van der Waals, *Z. Phys. Chem.* **13**, 657 (1894).

3. V. Ginzburg and L. Landau, *Zh. Eksp. Teor. Fiz.* **20**, 1064 (1950).

4. J. W. Cahn and J. E. Hilliard, *J. Chem. Phys.* **28**, 258 (1958).

5. N. D. Mermin, *Phys. Rev.* **137**, A1441 (1965).

6. T. V. Ramakrishnan and M. Yussouff, *Phys. Rev. B* **19**, 2775 (1979).

7. C. Ebner, W. F. Saam, and D. Stroud, *Phys. Rev. A* **14**, 2264 (1976).

8. W. F. Saam and C. Ebner, *Phys. Rev. A* **15**, 2566 (1977).

9. J. K. Percus, *J. Stat. Phys.* **15**, 505 (1976).

10. H. Reiss, H. L. Frisch, and J. L. Lebowitz, *J. Chem. Phys.* **31**, 369 (1959).

11. E. Helfand, H. Reiss, H. L. Frisch, and J. L. Lebowitz, *J. Chem. Phys.* **33**, 1379 (1960).

12. H. Reiss and R. V. Casberg, *J. Chem. Phys.* **61**, 1107 (1974).

13. R. Evans, *Adv. Phys.* **28**, 143 (1979).

14. R. Evans, in *Fundamentals of Inhomogeneous Fluids*, D. Henderson, editor, Marcel Dekker, New York, 1992.

15. M. Baus and J. F. Lutsko, *Physica A* **176**, 28 (1991).

16. H. Löwen, *Phys. Rep.* **237**, 249 (1994).

17. J. Wu, *AIChE J.* **52**, 1169 (2006).

18. S. L. Singh and Y. Singh, *Euro. Phys. Lett.* **88**, 16005 (2009).

19. P. Chaudhuri, S. Karmakar, C. Dasgupta, H. R. Krishnamurthy, and A. K. Sood, *Phys. Rev. Lett.* **95**, 248301 (2005).

20. S. van Teeffelen, C. N. Likos, N. Hoffmann, and H. Löwen, *Europhys. Lett.* **75**, 583 (2006).

21. J.-P. Hansen and I. McDonald, *Theory of Simple Liquids*, Academic Press, San Diego, CA, 1986.

22. T. Frankel, *The Geometry of Physics*, Cambridge University Press, Cambridge, UK, 1997.

23. J. K. Percus, in *The Equilibrium Theory of Classical Fluids*, H. L. Frisch and H. L. Lebowitz, editors, Benjamin, New York, 1964.

24. J. K. Percus, *Phys. Rev. Lett.* **8**, 462 (1962).
25. J. F. Lutsko and M. Baus, *Phys. Rev. A* **41**, 6647 (1990).
26. M. Baus and J. L. Colot, *Mol. Phys.* **56**, 804 (1985).
27. M. Baus, J. L. Colot, and H. Xu, *Mol. Phys.* **57**, 809 (1986).
28. B. Groh and B. Mulder, *Phys. Rev. E* **61**, 3811 (2000).
29. J. F. Lutsko, *Physica A* **366**, 229 (2006).
30. A. D. J. Haymet and D. W. Oxtoby, *J. Chem. Phys.* **74**, 2559 (1981).
31. D. W. Oxtoby and A. D. J. Haymet, *J. Chem. Phys.* **76**, 6262 (1982).
32. H. Löwen, T. Beier, and H. Wagner, *Europhys. Lett.* **9**, 791 (1989).
33. H. Löwen, T. Beier, and H. Wagner, *Z. Phys. B* **79**, 109 (1990).
34. P. Tarazona, *Phys. Rev. E* **47**, 4284 (1993).
35. J.-J. Weis, *Mol. Phys.* **28**, 187 (1974).
36. W. A. Curtin, *J. Chem. Phys.* **88**, 7050 (1988).
37. M. Baus and J. L. Colot, *Mol. Phys.* **55**, 653 (1985).
38. M. Baus, *J. Phys.: Cond. Matt.* **1**, 3131 (1989).
39. J. F. Lutsko and M. Baus, *Phys. Rev. Lett.* **64**, 761 (1990).
40. W. A. Curtin and N. W. Ashcroft, *Phys. Rev. A* **32**, 2909 (1985).
41. A. R. Denton and N. W. Ashcroft, *Phys. Rev A* **39**, 4701 (1985).
42. J. F. Lutsko, *J. Chem. Phys.* **128**, 184711 (2008).
43. P. Tarazona, *Phys. Rev. A* **31**, 2672 (1985).
44. D. Hendersen and E. W. Grundke, *J. Chem. Phys.* **63**, 601 (1975).
45. M. Baus and J. L. Colot, *Phys. Rev. A* **36**, 3912 (1987).
46. W. A. Curtin and K. Runge, *Phys. Rev. A* **35**, 4755 (1987).
47. C. F. Tejero, *Phys. Rev. E* **55**, 3720 (1997).
48. C. N. Likos and N. W. Ashcroft, *Phys. Rev. E* **52**, 5714 (1995).
49. J. L. Barrat, J. P. Hansen, G. Pastore, and F. M. Waisman, *J. Chem. Phys.* **86**, 6360 (1987).
50. F. J. Rogers and D. A. Young, *Phys. Rev. A* **30**, 999 (1984).
51. Y. Rosenfeld and N. W. Ashcroft, *Phys. Rev. A* **20**, 1208 (1979).
52. W. G. Hoover, D. A. Young, and R. Grover, *J. Chem. Phys.* **56**, 2207 (1972).
53. B. B. Laird and D. M. Kroll, *Phys. Rev. A* **42**, 4810 (1990).
54. A. de Kuijper, W. L. Vos, J.-L. Barrat, J.-P. Hansen, and J. A. Schouten, *J. Chem. Phys.* **93**, 5187 (1990).
55. A. D. J. Haymet, in *Fundamentals of Inhomogeneous Fluids*, D. Henderson, ed., Marcel Dekker, New York, 1992.
56. D. C. Wang and A. P. Gast, *J. Chem. Phys.* **110**, 2522 (1999).
57. W. G. Hoover and F. M. Ree, *J. Chem. Phys.* **49**, 3609 (1968).
58. W. A. Curtin, *Phys. Rev. B.* **39**, 6775 (1989).
59. R. Ohnesorge, H. Löwen, and H. Wagner, *Phys. Rev. A* **43**, 2870 (1991).
60. R. Ohnesorge, H. Löwen, and H. Wagner, *Phys. Rev. E* **50**, 4801 (1994).
61. D. W. Marr and A. P. Gast, *Phys. Rev. E* **47**, 1212 (1993).
62. A. Kyrlidis and R. A. Brown, *Phys. Rev. E* **51**, 5832 (1995).
63. W. A. Curtin and N. W. Ashcroft, *Phys. Rev. Lett.* **59**, 2385 (1987).

64. A. R. Denton and N. W. Ashcroft, *Phys. Rev. A* **39**, 426 (1989).
65. S.-C. Kim and G. L. Jones, *Phys. Rev. A* **41**, 2222 (1990).
66. J. A. White and R. Evans, *J. Phys. Condens. Matter* **2**, 2435 (1990).
67. A. R. Denton and N. W. Ashcroft, *Phys. Rev. A* **44**, 1219 (1991).
68. A. R. Denton and N. W. Ashcroft, *Phys. Rev. A* **41**, 2224 (1990).
69. J. L. Barrat, J. P. Hansen, and G. Pastore, *Mol. Phys.* **63**, 747 (1988).
70. Z. W. Salsburg, R. W. Zwanzig, and J. G. Kirkwood, *J. Chem. Phys.* **21**, 1098 (1953).
71. T. K. Vanderlick, H. T. Davis, and J. K. Percus, *J. Chem. Phys.* **91**, 7136 (1989).
72. T. K. Vanderlick, H. T. Davis, and J. K. Percus, *J. Chem. Phys.* **91**, 7136 (1989).
73. J. K. Percus, *J. Chem. Phys.* **75**, 1316 (1981).
74. R. Roth, R. Evans, A. Lang, and G. Kahl, *J. Phys. Condens. Matter* **14**, 12063 (2002).
75. Y. Rosenfeld, *Phys. Rev. Lett.* **63**, 980 (1989).
76. Y. Rosenfeld, D. Levesque, and J.-J. Weis, *J. Chem. Phys.* **92**, 6818 (1990).
77. E. Kierlik and M. L. Rosinberg, *Phys. Rev. A* **42**, 3382 (1990).
78. S. Phan, E. Kierlik, M. L. Rosinberg, B. Bildstein, and G. Kahl, *Phys. Rev. E* **48**, 618 (1993).
79. E. Kierlik and M. L. Rosinberg, *Phys. Rev. A* **44**, 5025 (1991).
80. Y. Rosenfeld, M. Schmidt, H. Löwen, and P. Tarazona, *Phys. Rev. E* **55**, 4245 (1997).
81. J. F. Lutsko, *Phys. Rev. E* **74**, 021121 (2006).
82. P. Tarazona, *Physica A* **306**, 243 (2002).
83. P. Tarazona, *Phys. Rev. Lett.* **84**, 694 (2000).
84. J. A. Cuesta, Y. Martinez-Raton, and P. Tarazona, *J. Phys. Condens. Matter* **14**, 11965 (2002).
85. G. A. Mansoori, N. F. Carnahan, K. E. Starling, and T. W. Leland, Jr., *J. Chem. Phys.* **54**, 1523 (1971).
86. V. B. Warshavsky and X. Song, *Phys. Rev. E* **73**, 031110 (2006).
87. J. F. Lutsko, *Phys. Rev. E* **74**, 021603 (2006).
88. Y. Rosenfeld, *Phys. Rev. E* **50**, R3318 (1994).
89. J. A. Cuesta, *Phys. Rev. Lett.* **76**, 3742 (1996).
90. J. A. Cuesta and Y. Martínez-Ratón, *Phys. Rev. Lett.* **78**, 3681 (1997).
91. L. Lafuente and J. A. Cuesta, *Phys. Rev. Lett.* **93**, 130603 (2004).
92. Y. Martinez-Raton, J. A. Capitan, and J. A. Cuesta, *Phys. Rev. E* **77**, 051205 (2008).
93. M. Schmidt, *Phys. Rev. E* **60**, R6291 (1999).
94. M. Schmidt, *Phys. Rev. E* **62**, 4976 (2000).
95. C. N. Likos, A. Lang, M. Watzlawek, and H. Löwen, *Phys. Rev. E* **63**, 031206 (2001).
96. J. A. Barker, and D. Henderson, *J. Chem. Phys.* **47**, 4714 (1967).
97. D. Chandler and J. D. Weeks, *Phys. Rev. Lett.* **25**, 149 (1970).
98. D. Chandler, J. D. Weeks, and H. C. Andersen, *J. Chem. Phys.* **54**, 5237 (1971).
99. H. C. Andersen, D. Chandler, and J. D. Weeks, *Phys. Rev. A* **4**, 1597 (1971).
100. F. Lado, *Mol. Phys.* **52**, 871 (1984).
101. J. M. Kincaid and J. J. Weis, *Mol. Phys.* **34**, 931 (1977).
102. C. Rascón, L. Mederos, and G. Navascués, *Phys. Rev. E* **54**, 1261 (1996).
103. C. Rascón, L. Mederos, and G. Navascués, *J. Chem. Phys.* **105**, 10527 (1996).
104. V. B. Warshavsky and X. Song, *Phys. Rev. E* **69**, 061113 (2004).

105. P. Paricaud, *J. Chem. Phys.* **124**, 154505 (2006).
106. Y. Rosenfeld, *J. Chem. Phys.* **98**, 8126 (1993).
107. Y. Tang and J. Wu, *Phys. Rev. E* **70**, 011201 (2004).
108. S. Sokolowski and J. Fischer, *J. Chem. Phys.* **96**, 5441 (1992).
109. T. Wadewitz and J. Winkelmann, *J. Chem. Phys.* **113**, 2447 (2000).
110. Z. Tang, L. E. Scriven, and H. T. Davis, *J. Chem. Phys.* **95**, 2659 (1991).
111. G. Stell and O. Penrose, *Phys. Rev. Lett.* **51**, 1397 (1983).
112. W. A. Curtin and N. W. Ashcroft, *Phys. Rev. Lett.* **56**, 2775 (1986).
113. L. Verlet and D. Levesque, *Physica* **36**, 245 (1967).
114. J. P. Hansen and L. Verlet, *Phys. Rev.* **184**, 151 (1969).
115. F. H. Ree, *J. Chem. Phys.* **64**, 4601 (1976).
116. H. S. Kang, C. S. Lee, T. Ree, and F. H. Ree, *J. Chem. Phys.* **82**, 414 (1985).
117. B. Q. Lu, R. Evans, and M. M. Telo da Gamma, *Mol. Phys.* **55**, 1319 (1985).
118. J. F. Lutsko and G. Nicolis, *Phys. Rev. Lett.* **96**, 046102 (2006).
119. Y. Tang, *J. Chem. Phys.* **118**, 4140 (2003).
120. J. F. Lutsko, *J. Chem. Phys.* **127**, 054701 (2007).
121. M. Llano-Restrepo and W. G. Chapman, *J. Chem. Phys.* **97**, 2046 (1992).
122. P. Grosfils and J. F. Lutsko, *J. Chem. Phys.* **130**, 054703 (2009).
123. P. R. ten Wolde and D. Frenkel, *Science* **77**, 1975 (1997).
124. P. Grosfils and J. F. Lutsko, unpublished (2009).
125. F. van Swol and J. R. Henderson, *Phys. Rev. A* **40**, 2567 (1989).
126. D. Henderson, ed., *Fundamentals of Inhomogeneous Fluids*, Marcel Dekker, New York, 1992.
127. W. A. Steele, *Surf. Sci.* **36**, 317 (1973).
128. W. A. Steele, *The Interaction of Gases with Solid Surfaces*, Pergamon, Oxford, 1974.
129. A. J. Archer, D. Pini, R. Evans, and L. Reatto, *J. Chem. Phys.* **126**, 014104 (2007).
130. S. R. de Groot and P. Mazur, *Non-Equilibrium Thermodynamics*, Dover, New York, 1984.
131. J. W. Cahn, *J. Chem. Phys.* **42**, 93 (1965).
132. P. C. Hohenberg and B. I. Halperin, *Rev. Mod. Phys.* **49**, 435 (1977).
133. T. Munakata, *J. Phys. Soc. Japan* **43**, 1723 (1977).
134. T. Munakata, *J. Phys. Soc. Japan* **45**, 749 (1978).
135. B. Bagchi, *Physica A* **145**, 273 (1987).
136. W. Dieterich, H. L. Frisch, and A. Majhofer, *Z. Phys. B* **78**, 317 (1990).
137. K. Kawasaki, *J. Stat. Phys.* **93**, 527 (1998).
138. U. M. B. Marconi and P. Tarazona, *J. Chem. Phys.* **110**, 8032 (1999).
139. S. P. Das, *Rev. Mod. Phys.* **76**, 785 (2004).
140. T. R. Kirkpatrick and P. G. Wolynes, *Phys. Rev. A* **35**, 3072 (1987).
141. G.K.-L. Chan and R. Finken, *Phys. Rev. Lett.* **94**, 183001 (2005).
142. J. A. McLennan, *Introduction to Nonequilibrium Statsitical Mechanics*, Prentice-Hall, Englewood Cliffs, NJ, 1989.
143. A. J. Archer and M. Rauscher, *J. Phys. A Math. Gen.* **37**, 9325 (2004).
144. R. Zwanzig, *Phys. Rev.* **124**, 983 (1961).
145. H. Mori and H. Fujisaka, *Prog. Theor. Phys.* **49**, 764 (1973).

146. H. Mori, H. Fujisaka, and H. Shigematsu, *Prog. Theor. Phys.* **51**, 109 (1974).

147. S. Sinha and M. C. Marchetti, *Phys. Rev. A* **46**, 4942 (1992).

148. S.-K. Ma and G. F. Mazenko, *Phys. Rev. B* **11**, 4077 (1975).

149. T. R. Kirkpatrick and J. C. Nieuwoudt, *Phys. Rev. A* **33**, 2651 (1986).

150. T. R. Kirkpatrick and J. C. Nieuwoudt, *Phys. Rev. A* **33**, 2658 (1986).

151. U. M. B. Marconi and P. Tarazona, *J. Chem. Phys.* **110**, 8032 (1999).

152. C. W. Gardiner, *Handbook of Stochastic Methods*, Springer-Verlag, Berlin, 2004.

153. A. J. Archer and R. Evans, *J. Chem. Phys.* **121**, 4246 (2004).

154. A. J. Archer *J. Chem. Phys.* **130**, 014509 (2009).

155. H. Löwen, *J. Phys. Cond. Matter* **15**, V1 (2003).

156. H. Löwen, *J. Phys. Cond. Matter* **15**, V1 (2003).

157. J. Zinn-Justin, *Quantum Field Theory and Critical Phenomena*, Clarendon Press, Oxford, 2002.

158. D. Reguera and H. Reiss, *J. Chem. Phys.* **120**, 2558 (2004).

159. D. N. Zubarev, ed., *Nonequilibrium Statistical Thermodynamics*, Consultants Bureau, New York, 1974.

160. A. J. Archer, P. Hopkins, and M. Schmidt, *Phys. Rev. E* **75**, 040501 (2007).

161. J.G.E.M. Fraaije, *J. Chem. Phys.* **99**, 9202 (1993).

162. J.G.E.M. Fraaije, *J. Chem. Phys.* **100**, 6984 (1994).

163. J.G.E.M. Fraaije, B. A. C. van Vlimmeren, N. M. Maurits, M. Postma, O. A. Evers, C. Hoffmann, P. Altevogt, and G. Goldbeck-Wood *J. Chem. Phys.* **106**, 4260 (1997).

164. J. Dzubiella and C. N. Likos, *J. Phys. Cond. Matter* **15**, L147 (2003).

165. S. van Teeffelen, C. N. Likos, and H. Löwen, *Phys. Rev. Lett.* **100**, 108302 (2008).

166. M. Rex and H. Löwen, *Phys. Rev. Lett.* **101**, 148302 (2008).

167. P. G. Vekilov, *Cryst. Growth Des.* **4**, 671 (2004).

168. M. J. Uline and D. S. Corti, *Phys. Rev. Lett.* **99**, 076102 (2007).

169. J. F. Lutsko, *Europhys. Lett.* **83**, 46007 (2008).

170. J. F. Lutsko, *J. Chem. Phys.* **129**, 244501 (2008).

171. J. W. Cahn and J. E. Hilliard, *J. Chem. Phys.* **31**, 688 (1959).

172. D. W. Oxtoby and R. Evans, *J. Chem. Phys.* **89**, 7521 (1988).

173. X. C. Zeng and D. W. Oxtoby, *J. Chem. Phys.* **94**, 4472 (1991).

174. V. Talanquer and D. W. Oxtoby, *J. Chem. Phys.* **100**, 5190 (1994).

175. D. Wales, *Energy Landscapes*, Cambridge University Press, Cambridge, 2003.

176. G. Henkelman and H. Jónsson, *J. Chem. Phys.* **113**, 9978 (2000).

177. E. Weinan, W. Ren, and E. V. Eijnden, *J. Chem. Phys.* **126**, (2007).

178. C. Qiu, T. Qian, and W. Ren, *J. Chem. Phys.* **129**, 154711 (2008).

179. J. K. Percus, *J. Stat. Phys.* **28**, 67 (1981).

NONADIABATIC CHEMICAL DYNAMICS IN INTERMEDIATE AND INTENSE LASER FIELDS

KAZUO TAKATSUKA AND TAKEHIRO YONEHARA

Department of Basic Science, Graduate School of Arts and Sciences, University of Tokyo, Komaba, 153-8902, Tokyo, Japan

CONTENTS

Advances in Chemical Physics, Volume 144, edited by Stuart A. Rice
Copyright © 2010 John Wiley & Sons, Inc.

I. INTRODUCTION

In this chapter, we study the effects of optical fields of intermediate and high intensity on the electron wavepacket dynamics that nonadiabatically couples with nuclear motions. In particular, we concentrate on an interplay between the intense laser fields and nuclear kinematic coupling of electrons due to the breakdown of the Born–Oppenheimer separation. Reflecting the recent theoretical progress of nonadiabatic transition, we first present an outline of our general yet practical theory of non-Born–Oppenheimer (or post-Born–Oppenheimer) theory for simultaneous determination of quantum electronic and nuclear wavepackets within the so-called on-the-fly scheme. In this theory, the nonadiabatic coupling is almost fully taken into account, irrespective of the functional form of avoided crossing and conical intersections. Even in a case where an obvious potential crossing is not present, the nonadiabatic coupling may play an important role, if the system is placed in an intense laser field, as discussed later.

Then, we introduce the classical vector potential of a laser field into the theory. In this particular chapter, however, we investigate the influence of the laser field on the above nonadiabatic interactions within the framework of the mean-field theory (using the so-called semiclassical Ehrenfest theory). Some very basic features of the nonadiabatic response of electrons to the laser fields are numerically presented. This is firstly because there is no report in the literature, to the best of our knowledge, about such *ab initio* nonadiabatic electron wavepacket dynamics of polyatomic (more than triatomic) molecules. Secondly, we are interested in laser manipulation of electronic states of molecules; that is, we are aiming at creating new electronic states that do not exist in nature as an eigenfunction of the molecular Hamiltonian. The ultimate goal of this manipulation is a control of chemical reactions.

The paper of Born and Oppenheimer published in 1927 [1] set the theoretical foundation of molecular framework, namely, the separation of electronic and nuclear motions on the basis of their significant difference in time and mass scales. The Born–Oppenheimer (BO) approximation is widely believed to be a very good approximation. So much so, people had not paid much attention how accurate or inaccurate the BO approximation is. Indeed, it is only recently that Takahashi and Takatsuka showed [2] that the error of the BO approximation is on the order of $(m/M)^{3/2}$, where m and M represent the masses of electron and nuclei, respectively. This explicitly indicates that the BO approximation is very good, typically to the order of 10^{-9} for $m/M \sim 10^{-6}$. However, once more than two potential energy surfaces come close to or cross each other, the perturbation expansion considered by Born and Oppenheimer [1] becomes invalid at the outset. Unfortunately, many of chemically interesting and important events take place in the vicinity of such "crossing" regions, where the nature of electronic states often qualitatively and dramatically change. Therefore, theory of nonadiabatic transition has been one of the central subjects in atomic and molecular science since the early 1930s [3–31]. (See also the reviews [32–35] for exhaustive lists of the relevant literature.)

However, the current status of chemical research is totally different from those in the age that had been dominated by the theories of Landau [3], Zener [4], and Stueckelberg [5]. For instance, for electron transfer dynamics induced by nonadiabatic coupling, whose time scale ranges several ten to several hundred attoseconds [36], *ab initio* calculation of electron wavepacket is now practically possible [37], and the main target in these studies is to track the electron wavepackets within and/or between molecules rather than the estimate of nonadiabatic transition probability. However, real-time track of electron wavepacket dynamics still demand new theoretical and algorithmic framework in practice.

Another very important factor that we should take account into the theory of electron wavepacket is the following two current aspects in laser technology. One is generation of very intense optical field provided to the extent of, say, 10^{16} W/cm^2, or more, which is comparable to the native Coulombic interactions among electrons and nuclei within a molecule. The other is applicability of an extremely short pulse laser, whose width is as narrow as down to the scale of 10 attoseconds (1 as $= 10^{-18}$ seconds) [38], which is already sufficiently comparable to the time scale of electronic motions in molecules [39–45]. These experimental progresses seem to be quite exciting and promising for molecular science in that it may reveal new properties of molecules and moreover may be utilized to assist to control the electronic structures as one wishes. The present study is also strongly motivated by the experimental progress.

Indeed, so many experimental and theoretical studies have been devoted to intense laser chemistry [20, 46–63]. Among others, the theoretical study on nonadiabatic effect and its indirect coupling with laser field in LiH and HF molecules at the single-determinant-electrons single-Gaussian-per-one-freedom-nuclei level by the group of Öhrn [46–48] is particularly relevant to the present work. As another important work in this field, Baer studied the electronic–nuclear coupled Hamiltonian under an external field in terms of the Born–Oppenheimer–Huang expansion [20]. The "semi-abstract form" used in their work is also related to the present work.

In the present chapter, we study *ab initio* nonadiabatic electron wavepacket dynamics in the presence of intense laser fields. Since the general solutions to the full (electronic and nuclear) Schrödinger equation are not available, except for very simple cases like hydrogen molecule cation [51, 64, 65], one needs a strategy to cope with nonadiabatic electron wavepacket dynamics, depending on what aspect of the target molecules is to be studied. One of the strong attentions in the attosecond electron dynamics is paid to multiple ionization followed by Coulomb explosion and the above threshold ionization (ATI) along with the relevant high-order harmonic generation (HHG). In the present chapter, however, we do not consider those ionization processes, but our concern is with the bound-state electron wavepacket dynamics only, which is supposed to have a direct relevance to chemical reaction dynamics. We will actually show in the later section that much nontrivial chemical dynamics is to be observed in this conjunction.

This chapter is structured as follows. The full non-Born–Oppenheimer electronic and nuclear wavepacket theory is first presented in Sections II and III for the dynamics of molecules in vacuum. On the other hand, in the case of laser field chemistry, which will be discussed in Section IV, we here take an approach of nonadiabatic coupling between full quantum electron dynamics and classical-like non-Born–Oppenheimer nuclear paths [66, 67]. This approach is justified as follows. For a short time dynamics, actually ranging from 10 attoseconds up to a hundred attoseconds, the nuclei should not move significantly and the associated quantum mechanical effect may be neglected. Besides, the wavelength of the matter wave of nuclei is very short anyway. Thus the classical-like approximation to nuclear motion must be a good approximation to start with. These paths can be eventually quantized in terms of semiclassical mechanics or the path-integral ansatz. In our future work, nuclear motion in laser field will be also quantized according to our general scheme of field-free non-Born–Oppenheimer dynamics of Section III. In Section V, we present numerical examples of nonadiabatic dynamics in intermediate and intense laser fields in some systematic manner, ranging from a basic examination of the effects of nonadiabatic coupling elements to a control of nonadiabatic transition in terms of lasers.

II. ELECTRONIC NONADIABATIC THEORY

We begin the theoretical part with the mixed quantum-classical representation of molecule, which is reduced to the semiclassical Ehrenfest theory (SET) by approximation. It is this approximation to which we introduce the vector potential of laser field in this chapter. SET is an approximate theory of nonadiabatic transition, which is not an exceptional situation but ubiquitous in many chemical reactions [35, 68, 69]. There are two basic ways to consider the kinematic coupling between electrons and nuclei, nonadiabatic interaction. Let us start to briefly review this aspect.

A. Nuclear Wavepacket Theory

Within the nonrelativistic scheme, the quantum-mechanical molecular Hamiltonian is written generally as

$$H(\mathbf{r}, \mathbf{R}) = \frac{1}{2} \sum_k \hat{P}_k^2 + H^{(\mathrm{el})}(\mathbf{r}; \mathbf{R}) \tag{1}$$

where many-body electronic Hamiltonian is defined by

$$H^{(\mathrm{el})}(\mathbf{r}; \mathbf{R}) = \frac{1}{2} \sum_j \hat{p}_j^2 + V_c(\mathbf{r}; \mathbf{R}) \tag{2}$$

Here, $\mathbf{r}$ and $\mathbf{R}$ denote the electronic and nuclear coordinates, respectively, and $\hat{p}_j$ and $\hat{P}_k$ are the operators of their conjugate momenta. $V_c(\mathbf{r}; \mathbf{R})$ is the Coulombic interaction potential among electrons and nuclei. The mass weighted coordinates are used throughout.

Because nuclei move far slower than electrons due to their heavy masses, it is the standard practice to expand the total wavefunction $\Psi(\mathbf{r}, \mathbf{R}, t)$ in static electronic basis functions $\{\Phi_I(\mathbf{r}; \mathbf{R})\}$, which are determined at each nuclear position

R, in such a way that

$$\Psi(\mathbf{r}, \mathbf{R}, t) = \sum_I \chi_I(\mathbf{R}, t)\Phi_I(\mathbf{r}; \mathbf{R}) \tag{3}$$

We assume that electronic basis set $\{|\Phi_I\rangle\}$, either adiabatic or diabatic, is orthonormal:

$$\langle \Phi_I(\mathbf{R})|\Phi_J(\mathbf{R})\rangle = \delta_{IJ} \tag{4}$$

In what follows, the bra–ket inner product represents integration over the electronic coordinates.

Inserting this form of the total wavefunction into time-dependent Schrödinger equation and using the above orthonormal property, one obtains [12, 20, 31, 68, 70, 71]

$$i\hbar\dot\chi_I = \frac{1}{2}\sum_k \hat{P}_k^2 \chi_I - i\hbar \sum_k \sum_J X_{IJ}^k \hat{P}_k \chi_J - \frac{\hbar^2}{2}\sum_k \sum_J Y_{IJ}^k \chi_J + \sum_J H_{IJ}^{(el)}\chi_J \tag{5}$$

where

$$X_{IJ}^k = \left\langle \Phi_I \left| \frac{\partial \Phi_J}{\partial R_k} \right. \right\rangle, \qquad Y_{IJ}^k = \left\langle \Phi_I \left| \frac{\partial^2 \Phi_J}{\partial R_k^2} \right. \right\rangle \tag{6}$$

and

$$H_{IJ}^{(el)}(\mathbf{R}) = \left\langle \Phi_I \left| H^{(el)} \right| \Phi_J \right\rangle \tag{7}$$

This multistate coupled Schrödinger equation is known to be cast into a compact matrix form as [35]

$$i\hbar\dot\chi_I = \sum_j \left[\frac{1}{2}\sum_k \left[(\mathbf{I}\hat{P}_k - i\hbar\mathbf{X}^k)^2 \right]_{IJ} + H_{IJ}^{(el)} \right]\chi_J \tag{8}$$

where $[\mathbf{I}]_{IJ} = \delta_{IJ}$ and $[\mathbf{X}^k]_{IJ} = X_{IJ}^k$.

The above classic theory is mathematically perfect, but the reduced equations of motion for nuclear wavepackets, Eq. (5), are never easy to solve. Moreover, in the experimental situations where attosecond electron dynamics is "directly" observed and/or intense electromagnetic field is applied, it is far more desirable for the electronic wavefunctions also to include the time coordinate t.

B. The Mean-Field Theory or the Semiclassical Ehrenfest Theory

The semiclassical Ehrenfest theory (SET) provides a good (but approximate) starting point to look at nonadiabatic transition from the view point of electron wavepacket dynamics. The total wavefunction is assumed to have the form

(unnormalized)

$$\Psi(\mathbf{r}, \mathbf{R}, t) = \sum_{\text{path}} \Phi(\mathbf{r}; \mathbf{R}_{\text{path}}(t)) \delta(\mathbf{R} - \mathbf{R}_{\text{path}}(t)) \tag{9}$$

where the electronic wavepacket propagated along a path $\mathbf{R}_{\text{path}}(t)$ is expanded as

$$\Phi(\mathbf{r}; \mathbf{R}_{\text{path}}(t)) = \sum_{I} C_I(t) \Phi_I(\mathbf{r}; \mathbf{R}_{\text{path}}) \tag{10}$$

with the coefficients being supposed to satisfy the electronic time-dependent Schrödinger equation

$$i\hbar \frac{\partial}{\partial t} C_I = \sum_{J} \left[H_{IJ}^{(\text{el})}(R_{\text{path}}(t)) - i\hbar \sum_{k} \dot{R}_k X_{IJ}^k(\mathbf{R}_{\text{path}}(t)) \right] C_J \tag{11}$$

The nuclear "classical" path carrying the electronic wavepacket is assumed to be driven by the mean force (as a mathematical analog of the Hellmann–Feynman force), that is,

$$
\begin{aligned}
\ddot{\mathbf{R}}_{\text{path}}(t) &= -\left\langle \Phi(\mathbf{R}_{\text{path}}(t)) \left| \frac{\partial H^{(\text{el})}}{\partial \mathbf{R}} \right| \Phi(\mathbf{R}_{\text{path}}(t)) \right\rangle \\
&= -\sum_{I} |C_I(t)|^2 \left\langle \Phi_I(\mathbf{R}_{\text{path}}) \left| \frac{\partial H^{(\text{el})}}{\partial \mathbf{R}} \right| \Phi_I(\mathbf{R}_{\text{path}}) \right\rangle
\end{aligned}
\tag{12}
$$

As seen above, SET is described on quite an intuitive basis, although it seemingly looks sound. We will give an explicit derivation of SET with correction terms later in this section. An advantage of SET is that it gives quite accurate transition amplitudes, as far as a single passage of the crossing region is concerned. On the other hand, an obvious drawback is, as repeatedly claimed, that the path thus driven is forced to run on an artificial potential energy function built by an average over the relevant adiabatic potential surfaces in the sense of Eq. (12).

C. Path Branching as the Essential Feature of Nonadiabatic Dynamics

The paths in the semiclassical Ehrenfest theory are not satisfactory not only because of the above wrong force but also because the individual pieces of the wavefunction in Eq. (9) propagated along a single path keep on it before and after the passage, that is,

$$\Phi(\mathbf{r}; \mathbf{R}_{\text{path}}(t_{\text{before}})) \delta(\mathbf{R} - \mathbf{R}_{\text{path}}(t_{\text{before}})) \rightarrow \Phi(\mathbf{r}; \mathbf{R}_{\text{path}}(t_{\text{after}})) \delta(\mathbf{R} - \mathbf{R}_{\text{path}}(t_{\text{after}}))$$

$$\tag{13}$$

There is no way for a state on a single path to give birth to the so-called quantum entanglement between the electronic and nuclear motions, which should be represented as, for instance,

$$\Phi(\mathbf{r}; \mathbf{R}_{\text{path}}(t_{\text{before}}))\delta(\mathbf{R}-\mathbf{R}_{\text{path}}(t_{\text{before}}))$$
$$\rightarrow \Phi_I(\mathbf{r}; \mathbf{R}_I(t_{\text{after}}))\delta(\mathbf{R}-\mathbf{R}_I(t_{\text{after}})) + \Phi_J(\mathbf{r}; \mathbf{R}_J(t_{\text{after}}))\delta(\mathbf{R}-\mathbf{R}_J(t_{\text{after}})) \tag{14}$$

where Φ_I and Φ_J create different potential energy surfaces. However, the mechanism of creating such a bifurcation in the total wavefunction (not only in the electronic part) should be a key feature to be considered for nonadiabatic dynamics.

To describe the total wavefunction based on ray solutions of nuclei, that is, $\mathbf{R}_I(t)$, we would rather expand the total wavefunction, in contrast to the classic form of Eq. (3), as

$$\Psi(\mathbf{r}, \mathbf{R}, t) = \sum_I^{\text{path}} c_I(t)\Phi_I(\mathbf{r}; \mathbf{R}_I(t))\chi_I(\mathbf{R}-\mathbf{R}_I(t), t) \tag{15}$$

where $\chi_I(\mathbf{R}-\mathbf{R}_I(t), t)$ is a (normalized) nuclear wavepacket localized nearby a path position $\mathbf{R}_I(t)$, while $\Phi_I(\mathbf{r}; \mathbf{R}_I(t))$ is a (normalized) electronic wavepacket at this point. One of our final goals is to build *ab initio* non-Born–Oppenheimer quantum chemistry, which enables us to track the total wavefunction within an on-the-fly scheme. Both nonadiabatic electron wavepacket and a sharp nuclear wavepacket are evolved in time along "classical" paths. To describe the entanglement situation, we impose the following condition:

$$\Phi(\mathbf{r}; \mathbf{R}_K(t_{\text{before}}))\chi(\mathbf{R}-\mathbf{R}_K(t_{\text{before}})) \rightarrow \sum_L^{\text{path}} c_L(t)\Phi_L(\mathbf{r}; \mathbf{R}_L(t_{\text{after}}))\chi_L(\mathbf{R}-\mathbf{R}_L(t_{\text{after}}))$$
$$\tag{16}$$

which is equivalent to a requirement that the individual paths should branch (bifurcate) to plural pieces.

Path branching is in general not realized in a well-posed initial value problem such as the ordinary differential equations like the Newtonian equations of motion giving classical trajectories. Therefore a mathematical trick is necessary to materialize branching. As an example, Takatsuka introduced a rotation of the Planck constant into the complex plane, thereby transforming the Hamiltonian so as to be anti-Hermitian [66]. Another tricky but very widely used technique is the so-called surface hopping model of Tully [15]. This method can represent the feature of path-branching partly, if used appropriately. However, we need a fundamental and consistent theory to determine naturally branching paths that at the same time couple with the electronic transition dynamics, thereby giving rise to an correct transition amplitude up to the phase.

D. A Generalized Classical Mechanics in Terms of Force Matrix to Generate Branching Paths

1. Basic Equations of Motion

To formulate the electron wavepacket dynamics on more clear basis, we begin by representing the total Hamiltonian operator in a basis set $\{|\Phi_I(\mathbf{R})\rangle|\mathbf{R}\rangle\}$ such that [66]

$$H(\mathbf{R}, \text{elec}) \equiv \frac{1}{2}\sum_k \left(\hat{P}_k - i\hbar \sum_{IJ} |\Phi_I\rangle X_{IJ}^k \langle \Phi_J| \right)^2 + \sum_{IJ} |\Phi_I\rangle H_{IJ}^{(\text{el})} \langle \Phi_J| \quad (17)$$

Notice a formal difference of this representation from Eq. (8). One can readily reduce it to the mixed quantum-classical representation by replacing the nuclear momentum operator $\hat{P}_k$ with its classical counterparts P_k such that

$$\tilde{H}(\mathbf{R}, \mathbf{P}, \text{elec}) \equiv \frac{1}{2}\sum_k \left(P_k - i\hbar \sum_{IJ} |\Phi_I\rangle X_{IJ}^k \langle \Phi_J| \right)^2 + \sum_{IJ} |\Phi_I\rangle H_{IJ}^{(\text{el})} \langle \Phi_J| \quad (18)$$

This is the starting point of our generalization of classical mechanics [67].

The dynamics of electron wavepackets is determined by the time-dependent variational principle as

$$\delta \int dt \langle \Phi(\mathbf{R}, t)| \left(i\hbar \frac{\partial}{\partial t} - \tilde{H}(\mathbf{R}, \mathbf{P}, \text{elec}) \right) |\Phi(\mathbf{R}, t)\rangle = 0 \quad (19)$$

(Both the left and right variation should be considered.) For an electron wavepacket expanded in a basis set as $|\Phi\rangle = \sum_I C_I(t)|\Phi_I\rangle$, this variational principle gives [67, 72]

$$i\hbar \frac{\partial}{\partial t} C_I = \sum_J \left[H_{IJ}^{(\text{el})} - i\hbar \sum_k \dot{R}_k X_{IJ}^k - \frac{\hbar^2}{4} \sum_k \left(Y_{IJ}^k + Y_{JI}^{k*} \right) \right] C_J \quad (20)$$

This expression looks similar to the standard representation of the semiclassical Ehrenfest theory [see Eq. (11)], except that Eq. (20) includes a nontrivial correction as the third term in its right-hand side.

With the above Hamiltonian, Eq. (18), an analogy to purely classical mechanics brings about the "canonical equations of motion" for nuclear classical variables $(\mathbf{R}, \mathbf{P})$ as [72]

$$\frac{d}{dt} R_k = \frac{\partial \tilde{H}}{\partial P_k} = P_k - i\hbar \sum_{IJ} |\Phi_I\rangle X_{IJ}^k \langle \Phi_J| \quad (21)$$

$$
\frac{d}{dt}P_k = -\frac{\partial \tilde{H}}{\partial R_k}
$$

$$
= i\hbar \sum_l \left(P_l - i\hbar \sum_{IJ} |\Phi_I\rangle X_{IJ}^l \langle \Phi_J|\right) \left(\sum_{KL} \frac{\partial (|\Phi_K\rangle X_{KL}^l \langle \Phi_L|)}{\partial R_k}\right)
$$

$$
- \frac{\partial}{\partial R_k} \left(\sum_{IJ} |\Phi_I\rangle H_{IJ}^{(el)} \langle \Phi_J|\right) \tag{22}
$$

As noticed immediately, the dynamics of $(\mathbf{R}, \mathbf{P})$ in these equations are essentially different from that of the purely classical counterparts in that the actions include the electronic ket–bra vectors. Rather, we should say more explicitly that the Hamiltonian Eq. (17), the force of Eq. (22), and the velocity of Eq. (21) are all defined in a joint space composed of the electronic Hilbert space and nuclear phase space. Therefore the quantities $\frac{d}{dt}R_k$ and $\frac{d}{dt}P_k$ work as transition operators for the electronic states. To emphasize this difference, we express our quantities with the calligraphic fonts as $\dot{\mathcal{R}}^k \equiv \frac{d}{dt}R_k$ and $\ddot{\mathcal{R}}^k \equiv \frac{d^2}{dt^2}R_k$. After some manipulation [67, 72], we find

$$
-\frac{d^2}{dt^2}\mathcal{R}^k = \sum_{IJ} \frac{\partial}{\partial R_k}(|\Phi_I\rangle H_{IJ}^{(el)} \langle \Phi_J|)
$$

$$
= -i\hbar \sum_{IJ} \sum_l \dot{R}_l \left[\frac{\partial}{\partial R_k}(|\Phi_I\rangle X_{IJ}^l \langle \Phi_J|) - \frac{\partial}{\partial R_l}(|\Phi_I\rangle X_{IJ}^k \langle \Phi_J|)\right] \tag{23}
$$

These "classical" operators can appear as physically meaningful quantities only after the electronic states are specified to be operated on. This implies that the force operators should be transformed to a matrix form by sandwiching two electronic states, say, $\langle \Phi_I|$ and $|\Phi_J\rangle$, as

$$
\mathcal{F}_{IJ}^k \equiv \left\langle \Phi_I \left| \ddot{\mathcal{R}}^k \right| \Phi_J \right\rangle
$$

$$
= -\sum_K \left[X_{IK}^k H_{KJ}^{(el)} - H_{IK}^{(el)} X_{KJ}^k\right] - \frac{\partial H_{IJ}^{(el)}}{\partial R_k} + i\hbar \sum_l \dot{R}_l \left[\frac{\partial X_{IJ}^l}{\partial R_k} - \frac{\partial X_{IJ}^k}{\partial R_l}\right] \tag{24}
$$

which is equivalent to

$$
\mathcal{F}_{IJ}^k = -\left(\frac{\partial H^{(el)}}{\partial R_k}\right)_{IJ} + i\hbar \sum_l \dot{R}_l \left[\frac{\partial X_{IJ}^l}{\partial R_k} - \frac{\partial X_{IJ}^k}{\partial R_l}\right] \tag{25}
$$

if the basis set used is complete. We refer to $\mathcal{F}_{IJ}^k$ as the force matrix. Obviously, the first term in the right-hand side of Eq. (25) comes from the electronic energy, while the second one represents the recoil from the delayed (kinematic) response of electrons to nuclear motions. This last term represents a force acting in the directions perpendicular to the velocity vector $\{\dot{R}_k\}$ as the magnetic Lorentz force does [67]. Therefore it appears only in the multidimensional theory. The mathematical origin of the Lorentz-like force is already suggested in the form of Eq. (17).

In the case of

$$X^k_{IJ} = 0 \tag{26}$$

it is most convenient to adopt the adiabatic wavefunctions $\{\psi_\alpha\}$, satisfying

$$H^{(el)}\psi_\alpha = E^{ad}_\alpha \psi_\alpha \tag{27}$$

Then we have

$$\mathcal{F}^k_{\alpha\beta} = -\delta_{\alpha\beta} \frac{\partial E^{ad}_\beta}{\partial R_k} \tag{28}$$

subject to the condition Eq. (26). Therefore the force matrix in the adiabatic limit reproduces the Born–Oppenheimer forces in its diagonal elements.

2. Reduction to the Semiclassical Ehrenfest Theory with Correction Terms

We here detour to see what is expected if we take an average of the force matrix in Eq. (24) over an electronic wavepacket $|\Phi\rangle = \sum_I C_I |\Phi_I\rangle$. It is immediate for us to obtain the mean-field force

$$
\begin{aligned}
\langle \Phi | \ddot{\mathcal{R}}^k | \Phi \rangle = & -\sum_{I,J,K} C^*_I \left(X^k_{IK} H^{(el)}_{KJ} - H^{(el)}_{IK} X^k_{KJ} \right) C_J - \sum_{IJ} C^*_I \frac{\partial H^{(el)}_{IJ}}{\partial R_k} C_J \\
& + i\hbar \sum_l \dot{R}_l \left[\left\langle \frac{\partial \Phi}{\partial R_k} \Big| \frac{\partial \Phi}{\partial R_l} \right\rangle - \left\langle \frac{\partial \Phi}{\partial R_l} \Big| \frac{\partial \Phi}{\partial R_k} \right\rangle \right]
\end{aligned}
\tag{29}
$$

If the basis set $\{\Phi_I\}$ were complete, the above equation reads

$$
\langle \Phi | \ddot{\mathcal{R}}^k | \Phi \rangle = -\left\langle \Phi \Big| \frac{\partial H^{(el)}}{\partial \mathbf{R}} \Big| \Phi \right\rangle + i\hbar \sum_l \dot{R}_l \left[\left\langle \frac{\partial \Phi}{\partial R_k} \Big| \frac{\partial \Phi}{\partial R_l} \right\rangle - \left\langle \frac{\partial \Phi}{\partial R_l} \Big| \frac{\partial \Phi}{\partial R_k} \right\rangle \right] \tag{30}
$$

which is equivalent to the semiclassical Ehrenfest theory except for the last summation in the right-hand side. This last sum over l, giving the Lorentz-like force, is a correction to the conventional semiclassical Ehrenfest theory. This term vanishes when Φ is real-valued and/or $\mathbf{R}$ is one-dimensional. However, one may neglect this term assuming that its magnitude is small because of the presence of $\hbar$. Note that we have used a finite basis representation of the force matrix, Eq. (24), because virtually in any case we cannot employ a totally complete basis [67, 72]. Thus the mean force has been formally derived. The derivation clearly uncovers that the mean-field theory is not always good, since it always takes an average over the entire wavepacket irrespective of wavepacket bifurcation.

3. The Eigenforces

The electronic wavepacket is to be carried along nuclear paths that are driven by the force matrix. Let us see how it is materialized. Suppose we have an electronic wavepacket $\Phi(\mathbf{r}; \mathbf{R}(t))$ at a phase-space point $(\mathbf{R}, \mathbf{P})$, where $\mathbf{R} = \mathbf{R}(t)$ and $\mathbf{P}$ is its conjugate momentum. To perform an electronic-state mixing among given basis functions $\{\Phi_I(\mathbf{r}; \mathbf{R})\}$ (either adiabatic or any diabatic basis), we first integrate Eq. (20) for a short time, say, Δt to give a new set of $\{C_I(t)\}$. Next we want to run a path using the force matrix $\mathcal{F}(\mathbf{R})$ again for a short time Δt. However, the presence of the off-diagonal elements in the force matrix can induce additional (unnecessary) electronic-state mixing. To avoid this additional mixing, we *diabatize* the electronic basis for this time interval by diagonalizing the force matrix at $\mathbf{R}$ such that

$$\mathbf{U}(\mathbf{R})\mathcal{F}(\mathbf{R})\mathbf{U}(\mathbf{R})^{-1} = \begin{pmatrix} f_1(\mathbf{R}) & 0 & \cdots \\ 0 & f_2(\mathbf{R}) & \\ \vdots & & \ddots \end{pmatrix} \tag{31}$$

with the associated electronic basis-set transformation

$$\mathbf{U}(\mathbf{R}) \begin{pmatrix} \Phi_1(\mathbf{r}; \mathbf{R}) \\ \Phi_2(\mathbf{r}; \mathbf{R}) \\ \vdots \end{pmatrix} = \begin{pmatrix} \lambda_1(\mathbf{r}; \mathbf{R}) \\ \lambda_2(\mathbf{r}; \mathbf{R}) \\ \vdots \end{pmatrix} \tag{32}$$

The electronic wavepacket obtained as above at $\mathbf{R}$ may be re-expanded in the eigenfunctions $\{\lambda_K(\mathbf{r}; \mathbf{R})\}$ such that

$$\Phi(\mathbf{r}; \mathbf{R}(t)) = \sum_K D_K(t)\lambda_K(\mathbf{r}; \mathbf{R}) \tag{33}$$

Then, each electronic component $D_K(t)\lambda_K(\mathbf{r}; \mathbf{R})$ is carried by its own path that is to be driven by the eigenforce f_K, without electronic mixing among $\{\lambda_L(\mathbf{r}; \mathbf{R})\}$, to reach a new point after Δt. [Note, however, that mixing among $\{\lambda_L(\mathbf{r}; \mathbf{R})\}$ can of course occur in the electron dynamics even in $\lambda_L(\mathbf{r}; \mathbf{R})$ representation of Eq. (20).] Different eigenforces make different paths even if they start from a single phase-space point $(\mathbf{R}, \mathbf{P})$ in such a way that

$$(\mathbf{R}, \mathbf{P}) \rightarrow (\mathbf{R}_K, \mathbf{P}_K) \tag{34}$$

Therefore a path at $\mathbf{R}$ is branched to as many pieces as the number of electronic states involved in the nonadiabatic coupling. The electronic-state mixing should be considered again at the individual points $(\mathbf{R}_K, \mathbf{P}_K)$, regarding the corresponding component $D_K(t)\lambda_K(\mathbf{r}; \mathbf{R})$ in Eq. (33) as the renewed condition to integrate Eq. (20). Again, the path branching is followed at each $(\mathbf{R}_K, \mathbf{P}_K)$. Hence, the cascade of path-branching should continue as long as the nonadiabatic coupling is present effectively. The full details of the entire procedure have been summarized elsewhere [67].

4. Using the Force Matrix (1): Phase-Space Averaging of Non-Born–Oppenheimer Paths

Tracking every branching path in the infinite cascade along with performing the electronic mixing along it would give an exact solution to the above dynamics of mixed quantum classical representation. It is obvious, however, that such a rigorous approach is extremely cumbersome and technically impossible, although we know that the branching feature reflects the essential physics behind nonadiabatic interaction. On the other hand, it is anticipated in the case where the coupling region is narrow enough that those branching paths should not geometrically deviate much from each other in phase space. In other words, they should localize along a representative path forming a tube-like structure. Therefore, we extract such a representative path by taking an average of phase-space points in the following manner:

(i) Suppose we have a path ending at $(\langle \mathbf{R}(t) \rangle, \langle \mathbf{P}(t) \rangle)$ in phase space. At this point, diagonalize the force matrix,

$$\mathcal{F}(\langle \mathbf{R} \rangle)|\lambda_K(\langle \mathbf{R} \rangle)) = |\lambda_K(\langle \mathbf{R} \rangle))f_K(\langle \mathbf{R} \rangle) \tag{35}$$

to obtain the eigenforces $\{f_K\}$ and its eigenstates $\{|\lambda_K\rangle\}$. The wavepacket $\Phi(\mathbf{r}; \langle \mathbf{R}(t) \rangle)$ is expanded in terms of these eigenstates as in Eq. (33).

(ii) The Kth eigenforce drives a path starting from $(\langle \mathbf{R}(t) \rangle, \langle \mathbf{P}(t) \rangle)$ for a short time Δt in terms of the Hamilton canonical equations of motion as

$$\mathbf{R}_K(t + \Delta t) = \langle \mathbf{R}(t) \rangle + \Delta \mathbf{R}_K \tag{36}$$

$$\mathbf{P}_K(t + \Delta t) = \langle \mathbf{P}(t) \rangle + \Delta \mathbf{P}_K \tag{37}$$

(iii) Now we average them into the form

$$\langle \mathbf{R}(t + \Delta t) \rangle = \langle \mathbf{R}(t) \rangle + \sum_K |D_K(t)|^2 \Delta \mathbf{R}_K \tag{38}$$

$$\langle \mathbf{P}(t + \Delta t) \rangle = \langle \mathbf{P}(t) \rangle + \sum_K |D_K(t)|^2 \Delta \mathbf{P}_K \tag{39}$$

which makes the next point $(\langle \mathbf{R}(t + \Delta t) \rangle, \langle \mathbf{P}(t + \Delta t) \rangle)$ of the representative path.

(iv) With this averaged point, we calculate

$$\mathcal{F}(\langle \mathbf{R}(t + \Delta t) \rangle) | \lambda_K(\langle \mathbf{R}(t + \Delta t) \rangle) \rangle = |\lambda_K(\langle \mathbf{R}(t + \Delta t) \rangle) \rangle f_K(\langle \mathbf{R}(t + \Delta t) \rangle)$$

and turn anew to step (ii).   (40)

Successive applications of the procedure (i)–(iii) gives a single finite path averaged in *phase space*. Let us recall that the semiclassical Ehrenfest theory demands to take an average over the electronic wavepacket to generate a single path. But, here in the present theory, many phase-space points are averaged to a single continuous path, which is not of course a classical path.

On the other hand, if the interacting region is not narrow, or if the strong nonadiabatic coupling brings about a large deviation among $(\Delta \mathbf{R}(t), \Delta \mathbf{P}(t))$, hard branching of paths is expected. In such cases, one may allow paths to branch at several places, compromising with additional computational tasks. We will show the relevant practice on this method later.

5.   Using the Force Matrix (2): Giving the Branching Paths

When a molecule gets into a coupling-free region after passing through an avoided-crossing zone, the force matrix smoothly becomes a diagonal matrix, whose diagonal elements represent the forces arising from the individual adiabatic potential energy surfaces. Also, the adiabatic wavefunctions become the eigenfunctions of the force matrix. At the same time, the mixing of electronic states is switched off gradually. Therefore, we should stop taking the average as in Eqs. (38) and (39) and simply let the individual components

$$D_K(t) \lambda_K(\mathbf{r}; \mathbf{R}_K(t))$$ (41)

run being driven by their own forces. Then the coefficients $D_K(t)$ coherently carry the information of transition amplitudes. In this way, a path can naturally branch as soon as the averaging is terminated.

Note that the averaging procedure of Eqs. (38) and (39) is possible even long after the nonadiabatic interaction is turned off, resulting in an unnecessary procedure to simply yield a spurious path. It is therefore practically crucial to choose the timing of natural branching. We will discuss these aspects later.

The branching paths have been introduced as an "classical" approximation of nuclear wavepackets in Eq. (15). Indeed, the examples of such branching paths are presented in our previous paper [72], which also highlights geometrical similarity between bifurcated quantum wavepackets and the corresponding branching paths.

Therefore, just as the individual pieces of interacting wavepackets do not have a constant energy, the electronic and nuclear energies along the individual pieces of branching paths are not rigorously conserved in the nonadiabatically coupling region.

The theory so far presented is general, and either adiabatic or any diabatic electronic wavefunctions are equally acceptable. In our practice below, however, we adopt the adiabatic representation and hence the following discussions proceed with $H_{KJ}^{(el)} = 0$ for $K \neq J$. In the nonadiabatic coupling region, a phase-space averaged path is created first, and it is relaxed to branch when the absolute value of the nonadiabatic coupling element is smaller than a predetermined value D, that is,

$$|\dot{R}X_{\alpha\beta}| \leq D \tag{42}$$

where α and β ($\alpha \neq \beta$) specify among the adiabatic states. Thus D is the only one parameter to dominate the path branching. As soon as Eq. (42) is found to be satisfied, we stop phase-space averaging of the paths. And then, using the relevant eigenforces, which are actually very close to the adiabatic forces already, we let the average path branch naturally [see Eqs. (38) and (39)], thus guiding to their adiabatic potential energy surfaces. These are so to say the path of the zeroth generation. Usually this single branching suffices to give a good result.

6. Using the Force Matrix (3): Further Branching

After the first branching performed as above, the individual paths are being driven by their own eigenforces. Suppose we are tracking one of them, say, the Kth path $(\mathbf{R}_K(t), \mathbf{P}_K(t))$. To emphasize that every force is generated along this path, we rewrite Eq. (31) explicitly as

$$\mathbf{U}(\mathbf{R}_K)\mathcal{F}(\mathbf{R}_K)\mathbf{U}(\mathbf{R}_K)^{-1} = \begin{pmatrix} f_1(\mathbf{R}_K) & 0 & \cdots \\ 0 & f_2(\mathbf{R}_K) & \\ \vdots & & \ddots \end{pmatrix} \tag{43}$$

where the dependence of the force matrix on $\mathbf{R}_K$ has been stressed. The right-hand side of this representation reminds us that other eigenforces—say, f_L—are also calculated along $\mathbf{R}_K(t)$. Therefore, at a point on the path $(\mathbf{R}_K(t), \mathbf{P}_K(t))$, one may switch the force from f_K to f_L to emanate *another* path, such that $(\mathbf{R}_K(t), \mathbf{P}_K(t)) \rightarrow (\mathbf{R}_L(t + \Delta t), \mathbf{P}_L(t + \Delta t))$ in terms of f_L. If one continues using f_K at the same point, it follows that $(\mathbf{R}_K(t), \mathbf{P}_K(t)) \rightarrow (\mathbf{R}_K(t + \Delta t), \mathbf{P}_K(t + \Delta t))$. Thus this procedure allows further branching. These are the paths of the first generation of branching. Along the paths of the first generation, we carry out the additional electronic-state mixing.

The individual branching paths of the first generation may branch again to just run toward the asymptotic region without further electronic mixing. Or, they can proceed to the second generation, taking another account of electronic-state mixing and the final branching of just releasing to the adiabatic surfaces. In a similar way, we can make a hierarchy (cascade) of branching. This is similar to the well-known procedure to take account of the so-called final-state interaction in dissociation dynamics.

The present method based on the phase-space averaging and natural branching is abbreviated as PSANB. The electronic transition amplitudes by PSANB has been numerically examined through comparison with the full quantum calculations based on the nuclear wavepacket approach and have been shown to be excellent [72].

III. NON-BORN–OPPENHEIMER QUANTUM CHEMISTRY

We have thus made up electron wavepacket dynamics nonadiabatically coupled with branching nuclear paths within the above quantum-classical mixed representation. The total wavefunction up to this point is explicitly written as follows. Suppose an electron wavepacket $\Phi_I(\mathbf{r}; \mathbf{R}(t_{\text{before}}))$ begins to run on an adiabatic potential energy surface I being carried by the associated nuclei, whose initial condition is $\mathbf{R}_{Ii}(t_{\text{before}})$, $\mathbf{P}_{Ii}(t_{\text{before}})$. Asymptotically, they are branched into pieces as

$$
\begin{aligned}
& \Phi_I(\mathbf{r}; \mathbf{R}(t_{\text{before}}))\delta(\mathbf{R}-\mathbf{R}_{Ii}(t_{\text{before}})) \\
& \to \sum_K \overset{\text{AES path}}{\sum_k} C_{Kk}(t_{\text{after}})\Phi_K(\mathbf{r}; \mathbf{R}_{Kk}(t_{\text{after}}))\delta(\mathbf{R}-\mathbf{R}_{Kk}(t_{\text{after}}))
\end{aligned}
\tag{44}
$$

where AES stands for adiabatic electronic state. The dependence of C_{Kk} on the momentum $\mathbf{P}_{Kk}(t_{\text{after}})$ is not explicitly expressed. This function fulfills the desired condition of Eq. (16) except that the nuclear wavefunction is not yet quantized. The remaining task is therefore to quantize the non-Born–Oppenheimer paths by replacing the delta function for the nuclei with an appropriate wavepacket.

A. Nuclear Wavepackets Along a Nonclassical Path

Then we would like to quantize the above-obtained non-Born–Oppenheimer paths by associating wavepackets, which correspond to $\chi_I(\mathbf{R}-\mathbf{R}_I(t), t)$ in our representation of Eq. (15). However, a serious technical difficulty we face is that they are not solutions of a single set of canonical equations of motion, although these paths are always characterized in phase space $(\mathbf{R}_I(t), \mathbf{P}_I(t))$. Hence, we cannot apply most of the convenient semiclassical theory to this particular problem, since they demand the information of the so-called stability matrix that is to be integrated along the classical trajectories (Newtonian paths). Therefore we apply our developed quantum wavepacket theory outlined below, which does not demand the stability matrix. The full detail of the theory can be found in Ref. 73.

1. Equation of Motion of a Nuclear Wavepacket

We begin with the following representation of a nuclear wavefunction on its coordinate q as

$$\Psi(q, t) = F(q, t)\exp\left(\frac{i}{\hbar}S(q, t)\right) \tag{45}$$

where S is assumed to satisfy the Hamilton–Jacobi equation, and thereby $\exp\left(\frac{i}{\hbar}S(q, t)\right)$ is regarded as a transformation function to determine $F(q, t)$, giving an equation of motion for it as

$$\frac{\partial F(q, t)}{\partial t} = \left(-\vec{v}\cdot\nabla - \frac{1}{2}(\nabla\cdot\vec{v})\right)F(q, t) + \frac{i\hbar}{2}\nabla^2 F(q, t) \tag{46}$$

where

$$\vec{v} = \nabla S(q, t) \tag{47}$$

which is the velocity. Equation (46) is readily rewritten in a manner of the Lagrangian view of fluid mechanics:

$$\begin{aligned}
\frac{dF(q, t)}{dt} &= \left(\frac{\partial}{\partial t} + \vec{v}\cdot\nabla\right)F(q, t) \\
&= -\frac{1}{2}(\nabla\cdot\vec{v})F(q, t) + \frac{i\hbar}{2}\nabla^2 F(q, t)
\end{aligned} \tag{48}$$

which suggests that this equation of motion can be integrated along a path in the velocity field $(q(t), v(t))$ [74, 75]. More explicitly, with the help of the Trotter decomposition [76], the short time propagation of $F(q, t)$ can be written as

$$\begin{aligned}
&F(q - q(t + \Delta t), t + \Delta t) \\
&\simeq \exp\left[\frac{i\hbar}{2}\Delta t\nabla^2\right]\exp\left[-\frac{1}{2}(\nabla\cdot\vec{v})\Delta t\right]F(q - q(t), t + \Delta t)
\end{aligned} \tag{49}$$

for a very small time step Δt.

2. Solving the Equation up to the Velocity Gradient Term

Among the two terms in Eq. (49), let us first consider the propagation by the velocity gradient term, that is,

$$\exp\left[-\frac{1}{2}(\nabla \cdot \vec{v})\Delta t\right] F(q-q(t), t+\Delta t) \tag{50}$$

which gives a semiclassical theory. The formal solution of Eq. (50) is given only in terms of the information of velocity field such that

$$F(q-q(t), t) = F(q-q(0), 0)\exp\left[-\int_0^t \frac{1}{2}(\nabla \cdot \vec{v})ds\right] \tag{51}$$

assuming that our treated problem does not contain the explicit dependence of time as in the molecular dynamics in a laser field.

The exponential function can be further simplified as

$$\int_0^t \frac{1}{2}(\nabla \cdot \vec{v})\,ds = \frac{1}{2}\int_0^t \sum_i \frac{\partial v_i}{\partial q_i}ds = \frac{1}{2}\sum_i \int_0^t \frac{ds}{dq_i}dv_i = \frac{1}{2}\sum_i \int_{v_i(0)}^{v_i(t)} \frac{dv_i}{v_i} \tag{52}$$

For one-dimensional case, this is just

$$\int_0^t \frac{1}{2}(\nabla \cdot \vec{v})\,ds = \frac{1}{2}\log v(t) - \frac{1}{2}\log v(0) \tag{53}$$

So, for a given time interval Δt, we have

$$F(q-q(t+\Delta t), t+\Delta t) = F(q-q(t), t)\left|\frac{v(t+\Delta t)}{v(t)}\right|^{-1/2}\exp\left[-i\pi\frac{M}{2}\right] \tag{54}$$

where M is the Maslov index in this sense.

3. Normalized Variable Gaussians

Next we should consider the term of quantum diffusion $\exp\left[\frac{i\hbar}{2}\Delta t\nabla^2\right]$ of Eq. (49), which essentially gives rise to the Feynman kernel for a free particle. The Feynman path summation is known to generally suffer from violent oscillations and lack of the integral measure. However, it is also well known that it can be rigorously and compactly treated for a Gaussian function. The price to utilize a Gaussian functions is, of course, that it is no longer an exact solution of Eq. (48) in general. Nevertheless, our numerical calculations shows that the Gaussian approximation to ADF gives accurate results. Therefore, we resume the dynamics of ADF with a Gaussian function reconsidering the velocity gradient term before the quantum diffusion is taken into account.

4. Rescaling the Exponents to the Gaussian Height

First of all, we recall that the norm of any ADF should be conserved, since

$$\int dq |\Psi(q)|^2 = \int dq \left| \exp\left[\frac{i}{\hbar} S(q,t) \right] F(q-q_k(t),t) \right|^2 = \int dq |F(q-q_k(t),t)|^2 = 1$$
(55)

where the suffix k specifies a path. We then assume the form of the Gaussian, but here its exponent is allowed to vary in such a way that

$$F(q-q_k(t),t) = \left(\frac{2}{\pi} \right)^{1/4} f(q_k(t)) \exp\left[-\gamma_k(t)(q-q_k(t))^2 \right]$$
(56)

Suppose the height of this function is multiplied by a positive factor R such that

$$F^{(R)}(q-q_k(t)) = RF^{(1)}(q-q_k(t))$$
$$= R \left(\frac{2}{\pi} \right)^{1/4} f(q_k(t)) \exp\left[-\gamma_k(t)(q-q_k(t))^2 \right]$$
(57)

Then, we want to renormalize this function by varying the exponent as

$$\int dq |F^{(R)}(q-q_k(t),t)|^2 = \int dq R^2 |F^{(1)}(q-q_k(t),t)|^2$$
$$= \sqrt{\frac{2}{\pi}} |f(q_k(t))|^2 \sqrt{\frac{\pi R^4}{2\mathrm{Re}(\gamma_k(t))}} = 1$$
(58)

Thus, if the exponent is rescaled as

$$\gamma_k(t) \rightarrow R^4 \gamma_k(t)$$
(59)

then this Gaussian is normalized again. We refer to this normalized Gaussian with variable exponent as ADF-NVG or simply NVG.

5. WKB Velocity-Gradient Term Revisited

In the propagation process of Eq. (49), we first perform the short time propagation with the velocity-gradient term, that is,

$$
\begin{aligned}
F^{\mathrm{vg}}(q-q(t+\Delta t)) &\equiv \exp\left[-\frac{1}{2}(\nabla\cdot\vec{v}_k)\Delta t\right]F(q-q_k(t+\Delta t)) \\
&= \left(\frac{2}{\pi}\right)^{1/4}\exp\left[-\frac{1}{2}(\nabla\cdot\vec{v}_k)\Delta t\right]f(q_k(t))\exp\left[-\gamma_k(t)(q-q_k(t))^2\right]
\end{aligned}
$$

(60)

Then, using the rescaling discussed above, we have

$$
F^{\mathrm{vg}}(q-q(t+\Delta t)) = \left(\frac{2}{\pi}\right)^{1/4}f^{\mathrm{vg}}(q_k(t+\Delta t))\exp\left[-\gamma_k^{\mathrm{vg}}(q-q_k(t+\Delta t))^2\right] \quad (61)
$$

where

$$
f^{\mathrm{vg}}(q_k(t+\Delta t)) = f(q_k(t))\left(\frac{v_k(t)}{v_k(t+\Delta t)}\right)^{1/2} \quad (62)
$$

and

$$
\gamma_k^{\mathrm{vg}}(t+\Delta t) = \gamma_k(t)\left(\frac{v_k(t)}{v_k(t+\Delta t)}\right)^2 \quad (63)
$$

6. Quantum Diffusion

We now apply the Feynman kernel to the above-obtained new Gaussian such that

$$
F(q-q_k(t+\Delta t),t) = \int dy K(q,y,\Delta t)F^{\mathrm{vg}}(y-q(t+\Delta t)) \quad (64)
$$

which simply gives

$$
\begin{aligned}
F(q-q_k(t+\Delta t),t) = \left(\frac{2}{\pi}\right)^{1/4}f(q_k(t))\left(\frac{v_k(t)}{v_k(t+\Delta t)}\frac{1/\gamma_k^{\mathrm{vg}}}{1/\gamma_k^{\mathrm{vg}}+i/a}\right)^{1/2} \\
\times \exp\left[-\frac{1}{1/\gamma_k^{\mathrm{vg}}+i/a}(q-q_k(t+\Delta t))^2\right]
\end{aligned}
$$

(65)

where

$$a = \frac{1}{2\hbar\Delta t} \tag{66}$$

which is regarded as a parameter characterizing the quantum diffusion for a short time interval Δt.

7. Integrated Equations of Motion

Thus ADF running on a point $(q(t), v(t))$ in the velocity field is explicitly approximated as above in terms of an NVG. Furthermore, ADF-NVG can be cast into more compact form allowing for a clear physical interpretation [73]. Write the NVG in the following form.

$$F(q-q_k(t), t) = \left(\frac{2}{\pi}\right)^{1/4} f_k(q_k(t)) \exp\left[-\frac{1}{c_k(t) + id_k(t)}(q-q_k(t))^2\right] \tag{67}$$

and we have

$$c_k(t+\Delta t) = c_k(t)\left(\frac{v_k(t+\Delta t)}{v_k(t)}\right)^2 \tag{68}$$

and

$$d_k(t+\Delta t) = d_k(t)\left(\frac{v_k(t+\Delta t)}{v_k(t)}\right)^2 + 2\hbar\Delta t \tag{69}$$

These are compactly integrated as

$$c_k(t) = \left(\frac{v_k(t)}{v_k(0)}\right)^2 c_k(0) \tag{70}$$

and

$$\frac{d_k(t)}{v_k(t)^2} = \frac{d_k(0)}{v_k(0)^2} + 2\hbar \int_0^t \frac{1}{v_k(s)^2}\, ds \tag{71}$$

The integral in the last term of Eq. (71) includes a singular integrand, which occurs when $v_k(s) = 0$. Nevertheless, this singularity can be removed and a finite integral is obtained [73]. This is a manifestation of quantum smoothing over (semi) classical singularity.

B. Combining Electronic Wavepacket Dynamics and Nuclear Wavepacket Theory

Now we can combine the theory of electron wavepackets on the non-Born–Oppenheimer branching paths (PSANB) and the nuclear wavepacket dynamics running along these paths (ADF-NVG). This is readily done as follows:

$$\Psi(\mathbf{r}, \mathbf{R}, t) = \sum_{K}^{\text{AES}} \sum_{k}^{\text{path}} C_{Kk}(t)\Phi_K(\mathbf{r}; \mathbf{R}_{Kk}(t))F_{\text{NVG}}(\mathbf{R}-\mathbf{R}_{Kk}(t), t) \qquad (72)$$

We thus have constructed a practical theory to describe the nonadiabatic total (electronic and nuclear) wavepackets for polyatomic molecules. As stated above, both the PSANB and ADF-NVG are numerically tested to give accurate results, and quite naturally therefore it turns out numerically that the above scheme of non-Born–Oppenheimer theory having the final form of Eq. (72) works quite well for nonadiabatic transition up to the phase factor [77].

IV. ELECTRONIC NONADIABATIC THEORY IN INTENSE LASER FIELD

Thus far, we have established non-Born–Oppenheimer theory for electronic and nuclear wavepacket dynamics, which actually gives rise to a practical *ab initio* methodology within the on-the-fly scheme. Therefore, the theory can be applied to a system in which electronic states are so highly degenerate that the notion of isolated adiabatic potential energy surface is less meaningful. However, this has been an achievement in chemical dynamics in vacuum. We next want to introduce the classical electromagnetic field into the theory. Unfortunately, however, electronic–nuclear quantum dynamics in intense laser field suffers from nontrivial difficulties. An obvious example is how to describe the wavepacket of ionizing electrons. We are still on the way to complete generalization of the above non-Born–Oppenheimer theory to laser-field chemical dynamics.

Indeed, most of the theoretical analyses so far made on existing experiments of intense laser chemistry, except for small molecules like H_2^+ and atoms [51, 64, 65], are based upon drastically simplified models. Also, studies with *ab initio*-level electronic-state calculations so far made have assumed fixed nuclei or complete neglect of nonadiabatic nuclear–electronic couplings. Besides, the laser field is mostly approximated in terms of the primitive electronic dipole moment multiplied by a sine-like oscillatory component and an envelope function.

In the following two sections, we study the effect of classical electromagnetic fields on nonadiabatic electron wavepacket dynamics. In view of the numerical difficulty that we have not yet overcome, we first formulate, in this section, the mean-field (semiclassical Ehrenfest) approach for a molecule embedded in laser field, which will make it possible to investigate the effect of laser on nonadiabatic electron wavepacket. Therefore this section is described in a self-contained manner, although many of the contents are closely correlated to those of the preceding two sections. Then the section of numerical studies follows, in which the theory is applied to several molecular systems to survey the interplay between nonadiabatic coupling and laser fields.

A.   The Total Hamiltonian in an Optical Field

We resume the discussion with the total Hamiltonian for a molecule that is placed in a classical electromagnetic field characterized in terms of the vector potential $\mathbf{A}$,

$$\mathcal{H}(\mathbf{r}, \mathbf{R}|\mathbf{A}) = \frac{1}{2}\sum_k \left(\hat{P}_k - \frac{q_k e}{c}A_k(\mathbf{R})\right)^2 + \mathcal{H}^{(\mathrm{el})}(\mathbf{r}; \mathbf{R}|\mathbf{A}) \qquad (73)$$

Here, $\mathbf{R} = \{R_k\}$ is the collective vector of all the nuclear positions with $\hat{\mathbf{P}} = \{\hat{P}_k\}$ being the vector of associated momenta, and $q_k e$ and c are the nuclear charge and the light velocity, respectively. The hat on $\hat{P}$ and so on denote the quantum operators. $\mathcal{H}^{(\mathrm{el})}(\mathbf{r}; \mathbf{R}|\mathbf{A})$ is the electronic Hamiltonian under the electromagnetic field

$$\mathcal{H}^{(\mathrm{el})}(\mathbf{r}; \mathbf{R}|\mathbf{A}) = \frac{1}{2}\sum_j \left(\hat{p}_j + \frac{e}{c}A_j(\mathbf{r})\right)^2 + V_c(\mathbf{r}; \mathbf{R}) \qquad (74)$$

where V_c collectively denotes the Coulombic interactions of electron–electron, electron–nuclei, and nuclei–nuclei. We denote $A_k(\mathbf{R})$ and $A_j(\mathbf{r})$ as the vector potentials for nuclei and electrons, respectively, unless confusion arises.

The time-dependent Schrödinger equation, $i\hbar\frac{\partial\Psi}{\partial t} = \mathcal{H}\Psi$, is reduced to coupled equations of motion for the so-called nuclear wavepackets $\chi_I(\mathbf{R}, t)$ such that

$$i\hbar\dot{\chi}_I = \frac{1}{2}\sum_k \left(\hat{P}_k - \frac{q_k e}{c}A_k\right)^2\chi_I - i\hbar\sum_k\sum_J X_{IJ}^k\left(\hat{P}_k - \frac{q_k e}{c}A_k\right)\chi_J - \frac{\hbar^2}{2}\sum_k\sum_J Y_{IJ}^k\chi_J$$

$$+ \sum_J^{(\mathrm{el})}\mathcal{H}_{IJ}^{(\mathrm{el})}(\mathbf{R}|\mathbf{A})\chi_J \qquad (75)$$

where

$$\mathcal{H}_{IJ}^{(\mathrm{el})}(\mathbf{R}|\mathbf{A}) = \left\langle \Phi_I \left| \left(\frac{1}{2}\sum_j^{\mathrm{el}}\left(\hat{p}_j + \frac{e}{c}A_j\right)^2 + V_c\right) \right| \Phi_J \right\rangle \qquad (76)$$

We rewrite $\mathcal{H}_{IJ}^{(\mathrm{el})}(\mathbf{R}|\mathbf{A})$ simply as $\mathcal{H}_{IJ}^{(\mathrm{el})}$ in what follows.

Equation (75) can be cast into a compact form as

$$i\hbar\dot{\chi}_I = \sum_J \left[\frac{1}{2}\sum_k \left[\left\{\mathbf{I}\left(\hat{P}_k - \frac{q_k e}{c}A_k\right) - i\hbar X^k\right\}^2\right]_{IJ} + \mathcal{H}_{IJ}^{(\mathrm{el})}\right]\chi_J \qquad (77)$$

where again $[\mathbf{I}]_{IJ} = \delta_{IJ}$ and $[\mathbf{X}^k]_{IJ} = X_{IJ}^k$.

B. Mixed Quantum Classical Representation

As we did in the field-free case, we here introduce the total Hamiltonian operator in the following representation using a basis set $\{|\Phi_I(\mathbf{R})\rangle|\mathbf{R}\rangle\}$ such that [78]

$$\mathcal{H}(\mathbf{R}, \text{elec}, \mathbf{A}) \equiv \frac{1}{2} \sum_k \left\{ \hat{P}_k - \frac{q_k e}{c} A_k - i\hbar \sum_{IJ} |\Phi_I\rangle X_{IJ}^k \langle \Phi_J| \right\}^2 + \sum_{IJ} |\Phi_I\rangle \mathcal{H}_{IJ}^{(el)} \langle \Phi_J|$$

(78)

This representation seems rather obvious (but not trivial) if compared with Eq. (73). (See also the similar Hamiltonian derived by Sun and Miller in the totally different context [12].) Again, the symbol elec in $\mathcal{H}(\mathbf{R}, \text{elec}, \mathbf{A})$ reminds that this Hamiltonian contains the electronic-state vectors explicitly. Here in Eq. (78), the nuclear momentum operators $\hat{P}_k$ are supposed to apply only on the nuclear wave function but not on the electronic-state vectors. This is because the nuclear derivative operators have been operated on all the other quantities such as $\Phi_I(\mathbf{R})$ in advance to this expression [recall Eq. (75)].

We next classicalize the above Hamiltonian (78) so as to access to its (approximate) solutions more easily. To do so, we construct a quantum-classical mixed representation of the Hamiltonian, $\tilde{\mathcal{H}}(\mathbf{R}, \mathbf{P}, \text{elec}, \mathbf{A})$, by simply exchanging the nuclear momentum operator $\hat{P}_k$ in $\mathcal{H}(\mathbf{R}, \text{elec}, \mathbf{A})$ with its classical counterpart P_k such that

$$\tilde{\mathcal{H}}(\mathbf{R}, \mathbf{P}, \text{elece}, \mathbf{A}) \equiv \frac{1}{2} \sum_k \left(P_k - \frac{q_k e}{c} A_k - i\hbar \sum_{IJ} |\Phi_I\rangle X_{IJ}^k \langle \Phi_J| \right)^2 + \sum_{IJ} |\Phi_I\rangle \mathcal{H}_{IJ}^{(el)} \langle \Phi_J|$$

(79)

Note that the treatment of the electronic states is the same as before. The tilde over $\mathcal{H}(\mathbf{R}, \mathbf{P}, \text{elec}, \mathbf{A})$ means the mixed quantum-classical representation. For the later convenience, we introduce the momentum shift operator Ω_k as

$$\Omega_k \equiv i\hbar \sum_{IJ} |\Phi_I\rangle X_{IJ}^k \langle \Phi_J| + \frac{q_k e}{c} A_k$$

(80)

giving the Hamiltonian a more compact form

$$\tilde{H}(\mathbf{R}, \mathbf{P}, \text{elec}, \mathbf{A}) = \frac{1}{2} \sum_k (P_k - \Omega_k)^2 + \sum_{IJ} |\Phi_I\rangle \mathcal{H}_{IJ}^{(el)} \langle \Phi_J|$$

(81)

which is next utilized to extract coupled equations of motion for constituent nuclei and electrons.

C. Coupled Dynamics of Electrons and Nuclei

1. Electrons

To track the dynamics of electron wavepackets coupled with the above nuclear motions in a self-consistent manner, we need the equations of motion for electrons. To derive them, we start from the time-dependent variational principle

$$\delta \int dt \langle \Phi(\mathbf{R}, t)| \left(i\hbar \frac{\partial}{\partial t} - \tilde{\mathcal{H}}(\mathbf{R}, \mathbf{P}, \text{elec}, \mathbf{A}) \right) |\Phi(\mathbf{R}, t)\rangle = 0 \qquad (82)$$

with $\tilde{\mathcal{H}}(\mathbf{R}, \mathbf{P}, \text{elec}, \mathbf{A})$ being the total Hamiltonian defined in Eq. (78), and here again we expand the electronic wavepacket as $|\Phi\rangle = \sum_I C_I(t)|\Phi_I\rangle$. Defining

$$\mathcal{G} \equiv \tilde{\mathcal{H}} - \frac{1}{2} \sum_k \left(P_k - \frac{q_k e}{c} A_k \right)^2 \qquad (83)$$

and utilizing its Hermiticity, $\mathcal{G}_{IJ} = \langle \Phi_I|\mathcal{G}|\Phi_J\rangle = (\mathcal{G}_{JI})^*$, we obtain

$$\text{Re} \sum_{IJ} \delta C_I^* \langle \Phi_I| \left(i\hbar \frac{\partial}{\partial t} - \mathcal{G} \right) |\Phi_J\rangle C_J = 0 \qquad (84)$$

The vector of the coefficients, $\{C_I\}$, which should satisfy

$$i\hbar \frac{\partial}{\partial t} C_I \equiv \sum_J \mathcal{G}_{IJ} C_J \qquad (85)$$

is one of the solutions of the variational equation, and these equations are exactly what we need. Taking account of the right variation also just as the left variation of Eq. (84), we obtain the coupled equations for the electron dynamics as

$$i\hbar \frac{\partial}{\partial t} C_I = \sum_J \left[\mathcal{H}_{IJ}^{(\text{el})} - i\hbar \sum_k \left(P_k - \frac{q_k e}{c} A_k \right) X_{IJ}^k - \frac{\hbar^2}{4} \sum_k (Y_{IJ}^k + Y_{JI}^{k*}) \right] C_J \qquad (86)$$

This is the correct generalization of the semiclassical Ehrenfest theory to dynamics in optical field [78].

It is interesting to see that the second derivative coupling elements Y_{IJ}^k in this expression stand independently from the external vector potential $\mathbf{A}$, whereas the first derivative coupling elements X_{IJ}^k are directly coupled with A_k. To be more precise, the second term on the right-hand side of Eq. (86) indicates that $X_{IJ}^k(\mathbf{R})$ mediates the interaction between the nuclear momentum P_k with the vector potential. More direct distortion of the electron wavepacket should be brought about by the vector potential included in $\mathcal{H}_{IJ}^{(\text{el})}$.

2. Nuclei

To establish the nuclear dynamics likewise, we apply the standard procedure of classical mechanics to the Hamiltonian (81) to obtain the canonical equations of nuclear motion in such a way that

$$\frac{d}{dt}R_k = \frac{\partial \tilde{\mathcal{H}}}{\partial P_k} = P_k - \Omega_k \tag{87}$$

$$\frac{d}{dt}P_k = -\frac{\partial \tilde{\mathcal{H}}}{\partial R_k} = \sum_l (P_l - \Omega_l)\frac{\partial \Omega_l}{\partial R_k} - \frac{\partial}{\partial R_k}\left(\sum_{IJ}|\Phi_I\rangle \mathcal{H}_{IJ}^{(el)}\langle\Phi_J|\right) \tag{88}$$

Here the actions of electrons on nuclei are represented jointly in the electronic Hilbert space and nuclear configuration space. Therefore, the quantities $\frac{d}{dt}R_k$ and $\frac{d}{dt}P_k$ should work as transition operators for the electronic states. To differentiate these quantities from their counterparts in the Newtonian mechanics, we express them with the calligraphic fonts as $\dot{\mathcal{R}}^k \equiv \frac{d}{dt}R_k$ and $\ddot{\mathcal{R}}^k \equiv \frac{d^2}{dt^2}R_k$.

Insertion of Eq. (22) into Eq. (21) leads to

$$\ddot{\mathcal{R}}^k = \frac{d^2}{dt^2}\mathcal{R}^k = \frac{d}{dt}P_k - \frac{d}{dt}\Omega_k$$

$$= (P - \Omega)\cdot\left(\frac{\partial\Omega}{\partial R_k}\right) - \frac{d}{dt}\Omega_k - \frac{\partial}{\partial R_k}\left(\sum_{IJ}|\Phi_I\rangle \mathcal{H}_{IJ}^{(el)}\langle\Phi_J|\right) \tag{89}$$

where $\mathbf{P} = (P_1, P_2, \ldots, P_N)$ and $\Omega = (\Omega_1, \Omega_2, \ldots, \Omega_N)$. Using Eq. (21) and

$$\frac{d}{dt}\Omega_k = \frac{\partial}{\partial t}\Omega_k + \dot{\mathbf{R}}\cdot\frac{\partial\Omega_k}{\partial\mathbf{R}} \tag{90}$$

with $\mathbf{R} = (R_1, \ldots, R_N)$, we obtain

$$\frac{d^2}{dt^2}\mathcal{R}^k = -\frac{\partial}{\partial R_k}\left(\sum_{IJ}|\Phi_I\rangle \mathcal{H}_{IJ}^{(el)}\langle\Phi_J|\right) - \frac{\partial\Omega_k}{\partial t} + \dot{\mathbf{R}}\cdot\left(\frac{\partial W}{\partial R_k} - \frac{\partial\Omega_k}{\partial\mathbf{R}}\right) \tag{91}$$

There are two notes for this equation: (i) The third term on the right-hand side serves as force acting perpendicular to the direction of $\dot{\mathbf{R}}$. (ii) It holds

$$\frac{\partial\Omega_k}{\partial t} = \frac{q_k e}{c}\frac{\partial A_k}{\partial t} \tag{92}$$

because only the term A_k in Ω_k explicitly depends on time t [recall Eq. (80)].

With use of the definition of Ω_k, Eq. (91) is further deformed as

$$
\frac{d^2}{dt^2}\mathcal{R}^k = -\sum_{IJ}\frac{\partial}{\partial R_k}(|\Phi_I\rangle\mathcal{H}_{IJ}^{(\mathrm{el})}\langle\Phi_J|)-\frac{q_k e}{c}\frac{\partial A_k}{\partial t}
$$
$$
+i\hbar\sum_{IJ}\sum_{l}\dot{R}_l\left[\frac{\partial}{\partial R_k}(|\Phi_I\rangle X_{IJ}^l\langle\Phi_J|)-\frac{\partial}{\partial R_l}(|\Phi_I\rangle X_{IJ}^k\langle\Phi_J|)\right] \quad (93)
$$
$$
+\frac{e}{c}\sum_{l}\dot{R}_l\left(\frac{\partial(q_l A_l)}{\partial R_k}-\frac{\partial(q_k A_k)}{\partial R_l}\right)
$$

To facilitate understanding how similar (different) this expression is to (from) the standard force in the classical electromagnetic field, we rewrite the suffices k and l in terms of those explicitly referring to the individual nuclei, like i_a, where a denotes a nucleus. Thus, i_a runs x_a, y_a, and z_a, which are, x, y, and z components for the nucleus a, respectively. Equation (23) is thus rewritten as

$$
\ddot{\mathcal{R}}^{i_a} = -\sum_{I,J}\frac{\partial}{\partial R_{i_a}}(|\Phi_I\rangle\mathcal{H}_{IJ}^{(\mathrm{el})}\langle\Phi_J|)
$$
$$
+i\hbar\sum_{I,J}\sum_{b}^{\mathrm{nuclei}}\sum_{j_b}\dot{R}_{j_b}\left[\frac{\partial}{\partial R_{i_a}}(|\Phi_I\rangle X_{IJ}^{j_b}\langle\Phi_J|)-\frac{\partial}{\partial R_{j_b}}(|\Phi_I\rangle X_{IJ}^{i_a}\langle\Phi_J|)\right] \quad (94)
$$
$$
+q_a e(\mathbf{E}_a+\dot{\mathbf{R}}_a\times\mathbf{B}_a)_{i_a}
$$

where the Maxwell equations, $\mathbf{E}_a = -\frac{1}{c}\frac{\partial\mathbf{A}_a}{\partial t}$ and $\mathbf{B}_a = \nabla_a\times\mathbf{A}_a$ have been used, and $q_a e$ is the charge of the ath nucleus. Note again that $\mathbf{A}_a$ depends only on (x_a, y_a, z_a) and t [79].

The force operators obtained above are readily transformed into a matrix form by specifying the electronic states we are interested in. For instance, for a given electronic configuration, say Φ_I, we have

$$
\langle\Phi_I|\ddot{\mathcal{R}}^{i_a}|\Phi_I\rangle = -\sum_{K}\left[X_{IK}^{i_a}\mathcal{H}_{KI}^{(\mathrm{el})}-\mathcal{H}_{IK}^{(\mathrm{el})}X_{KI}^{i_a}\right]-\frac{\partial\mathcal{H}_{II}^{(\mathrm{el})}}{\partial R_{i_a}}+q_a e(\mathbf{E}_a+\dot{\mathbf{R}}_a\times\mathbf{B}_a)_{i_a}
$$
$$
\quad (95)
$$

which turns out to be equivalent to

$$
\langle\Phi_I|\ddot{\mathcal{R}}^{i_a}|\Phi_I\rangle = -\left(\frac{\partial\mathcal{H}^{(\mathrm{el})}}{\partial R_{i_a}}\right)_{II}+q_a e(\mathbf{E}_a+\dot{\mathbf{R}}_a\times\mathbf{B}_a)_{i_a} \quad (96)
$$

only when the electronic basis is complete. Note the difference between

$$\frac{\partial \mathcal{H}_{IJ}^{(\mathrm{el})}}{\partial R_{i_a}} = \frac{\partial}{\partial R_{i_a}} \left\langle \Phi_I | \mathcal{H}^{(\mathrm{el})} | \Phi_J \right\rangle \tag{97}$$

and

$$\left(\frac{\partial \mathcal{H}^{(\mathrm{el})}}{\partial R_{i_a}} \right)_{IJ} = \left\langle \Phi_I \left| \frac{\partial \mathcal{H}^{(\mathrm{el})}}{\partial R_{i_a}} \right| \Phi_J \right\rangle \tag{98}$$

For the off-diagonal elements ($I \neq J$), we have

$$\left\langle \Phi_I \left| \ddot{\mathcal{R}}^{i_a} \right| \Phi_J \right\rangle$$

$$= -\sum_K \left[X_{IK}^{i_a} \mathcal{H}_{KJ}^{(\mathrm{el})} - \mathcal{H}_{IK}^{(\mathrm{el})} X_{KJ}^{i_a} \right] - \frac{\partial \mathcal{H}_{IJ}^{(\mathrm{el})}}{\partial R_{i_a}} + i\hbar \sum_b^{\mathrm{nuclei}} \sum_{j_b} \dot{R}_{j_b} \left[\frac{\partial X_{IJ}^{j_b}}{\partial R_{i_a}} - \frac{\partial X_{IJ}^{i_a}}{\partial R_{j_b}} \right] \tag{99}$$

or for the complete basis

$$\left\langle \Phi_I \left| \ddot{\mathcal{R}}^{i_a} \right| \Phi_J \right\rangle = -\left(\frac{\partial \mathcal{H}^{(\mathrm{el})}}{\partial R_{i_a}} \right)_{IJ} + i\hbar \sum_b^{\mathrm{nuclei}} \sum_{j_b} \dot{R}_{j_b} \left[\frac{\partial X_{IJ}^{j_b}}{\partial R_{i_a}} - \frac{\partial X_{IJ}^{i_a}}{\partial R_{j_b}} \right] \tag{100}$$

It thus turns out that the nonadiabatic force proportional to the velocity term $\dot{R}_{j_b}$, like the macroscopic friction, appears only in the off-diagonal elements. On the other hand, the magnetic Lorentz force appears only in the diagonal elements. Besides, due to the identity $X_{IJ}^k = -X_{JI}^k$ (assuming that all the basis functions are real), we observe

$$\left\langle \Phi_I \left| \ddot{\mathcal{R}}^{i_a} \right| \Phi_J \right\rangle = \left\langle \Phi_J \left| \ddot{\mathcal{R}}^{i_a} \right| \Phi_I \right\rangle^* \tag{101}$$

and hence the force matrix is a complex Hermitian matrix.

Finally, we recall $\mathcal{H}^{(\mathrm{el})} = \hat{H}^{(\mathrm{el})} + \hat{V}^{\mathrm{opt}}$, where

$$\hat{V}^{\mathrm{opt}}(\mathbf{r}) = \frac{1}{2} \sum_j^{\mathrm{electrons}} \left\{ \frac{e}{c} (\hat{p}_j A_j(\mathbf{r}) + A_j(\mathbf{r}) \hat{p}_j) + \frac{e^2}{c^2} A_j(\mathbf{r})^2 \right\} \tag{102}$$

and therefore it holds that

$$\frac{\partial \hat{V}^{\mathrm{opt}}(\mathbf{r})}{\partial R_{i_a}} = 0 \tag{103}$$

which in turn gives rise to

$$\frac{\partial \mathcal{H}^{(\mathrm{el})}}{\partial R_{i_a}} = \frac{\partial H^{(\mathrm{el})}}{\partial R_{i_a}} \tag{104}$$

where $H^{(\mathrm{el})}$ is the field-free electronic Hamiltonian. Thus, if the basis set were complete, we formally have [78]

$$\left\langle \Phi_I \middle| \ddot{\mathcal{R}}^{i_a} \middle| \Phi_J \right\rangle = -\left(\frac{\partial H^{(\mathrm{el})}}{\partial R_{i_a}}\right)_{IJ}$$
$$+ (1-\delta_{IJ}) i\hbar \sum_b^{\mathrm{nuclei}} \sum_{j_b} \dot{R}_{j_b} \left[\frac{\partial X_{IJ}^{j_b}}{\partial R_{i_a}} - \frac{\partial X_{IJ}^{i_a}}{\partial R_{j_b}}\right]$$
$$+ \delta_{IJ} q_a e (\mathbf{E}_a + \dot{\mathbf{R}}_a \times \mathbf{B}_a)_{i_a} \tag{105}$$

D. Mean-Field Approximation

1. Mean-Field Approximation to the Nuclear Dynamics

In spite of the rather intricate structure of the force matrix, Eq. (105), its average over a given wavepacket $|\Phi\rangle = \sum_I C_I |\Phi_I\rangle$—that is, the mean-field forces acting on nuclei—turn out to have such a strikingly simple form as

$$\langle \Phi | \ddot{\mathcal{R}}^{i_a} | \Phi \rangle = -\sum_{I,J,K} C_I^* \left(X_{IK}^{i_a} \mathcal{H}_{KJ}^{(\mathrm{el})} - \mathcal{H}_{IK}^{(\mathrm{el})} X_{KJ}^{i_a}\right) C_J - \sum_{IJ} C_I^* \frac{\partial \mathcal{H}_{IJ}^{(\mathrm{el})}}{\partial R_{i_a}} C_J$$
$$+ q_a e (\mathbf{E}_a + \dot{\mathbf{R}}_a \times \mathbf{B}_a)_{i_a} \tag{106}$$

This expression is theoretically valid even for finite bases. If a complete basis set was available, this expression is reduced to the form using the Hellmann–Feynman force as

$$\langle \Phi | \ddot{\mathcal{R}}^{i_a} | \Phi \rangle = -\sum_{I,J} C_I^* \left(\frac{\partial H^{(\mathrm{el})}}{\partial R_{i_a}}\right)_{IJ} C_J + q_a e (\mathbf{E}_a + \dot{\mathbf{R}}_a \times \mathbf{B}_a)_{i_a} \tag{107}$$

Further, if the optical interaction for electrons $\hat{V}^{\mathrm{opt}}(\mathbf{r})$ is independent of the nuclear coordinates as in Eq. (103), the Hellmann–Feynman force in Eq. (107) turns out to be free of the optical term such that

$$\langle \Phi | \ddot{\mathcal{R}}^{i_a} | \Phi \rangle = -\sum_{I,J} C_I^* \left(\frac{\partial H^{(\mathrm{el})}}{\partial R_{i_a}}\right)_{IJ} C_J + q_a e (\mathbf{E}_a + \dot{\mathbf{R}}_a \times \mathbf{B}_a)_{i_a} \tag{108}$$

2. Mean-Field Approximation to the Electron Wavepacket Dynamics

The mean-field approximation has an effect to the electronic wavepacket dynamics. Let us look back at Eq. (21) and take the average of $\dot{\mathcal{R}}^k$ over the wavepacket such that

$$\dot{R}_k \equiv \langle \Phi | \dot{\mathcal{R}}^k | \Phi \rangle = P_k - i\hbar \langle \Phi | \frac{\partial}{\partial R_k} | \Phi \rangle - \frac{q_k e}{c} A_k \qquad (109)$$

Therefore Eq. (86) is rewritten with the velocity vector as

$$i\hbar \frac{\partial}{\partial t} C_I = \sum_J \left[\mathcal{H}_{IJ}^{(el)} - i\hbar \sum_k \dot{R}_k X_{IJ}^k + \hbar^2 \sum_k \left(\langle \Phi | \frac{\partial}{\partial R_k} | \Phi \rangle - \frac{Y_{IJ}^k + Y_{JI}^{k*}}{4} \right) \right] C_J \qquad (110)$$

The last two terms of this equation are missing in the standard expression of the semiclassical Ehrenfest theory. Furthermore, only when the imaginary part of $\langle \Phi | \frac{\partial}{\partial R_k} | \Phi \rangle$ happens to be zero (note that the real part is identically zero), which is not necessarily the general case, we have

$$i\hbar \frac{\partial}{\partial t} C_I = \sum_J \left(\mathcal{H}_{IJ}^{(el)} - i\hbar \sum_k \dot{R}_k X_{IJ}^k - \frac{\hbar^2}{4} \sum_k (Y_{IJ}^k + Y_{JI}^{k*}) \right) C_J \qquad (111)$$

However, it is the usual practice to neglect the terms proportional to $\hbar^2$ under a belief that they are small.

E. Coupling Between the Nonadiabatic Force and the Laser Field

We next discuss the physical effects of the optical field in the presented formulation.

1. The Nonadiabatic Lorentz-like Force in Nuclear Dynamics

We have observed in Eq. (105) the two kinds of force acting on the nuclei in perpendicular directions to their path. One is the Lorentz force arising from the electromagnetic field, which appears only on the diagonal elements of the force matrix, Eq. (105). The other such force arises from the nonadiabatic coupling and is observed only in the off-diagonal elements. It should be noted that while the electromagnetic Lorentz force working on a nucleus a is generated within itself, the nonadiabatic Lorentz-like force is applied from other nuclei too. Note that this nonadiabatic Lorentz-like force is canceled away in the semiclassical Ehrenfest theory in the process of the electronic wavepacket averaging.

2. No Impulsive Force on the Nuclei Arising from the Electronic State Change

In looking at the force matrix [Eqs. (95), (99), and (105)], it is noticed that the force component arising from the distortion of the electron wavepacket through the change of the coefficients $\{C_I\}$ is formally missing in these expressions. This is because we have used time-independent (field independent) basis functions. On the other hand, the nuclear force in the mean-field theory, Eq. (107), is directly dependent on the change of the electronic state coefficients $\{C_I\}$. Therefore, in the force matrix formalism, there is no mechanism for the nuclei to feel an impulsive force arising from the sudden change of the electronic states.

3. Coupling Between the Nonadiabatic Force and the Lorentz Force

As noted above, the nuclear forces arising from the nonadiabatic coupling and those from the vector potential have appeared linearly in the additive form. This is because the nuclear force is obtained through the quantity $\frac{d}{dt}P_k$ and due to the linear relation of Eq. (80) to the momentum P_k. Nevertheless, we observe the coupling terms such as

$$X_{IK}^{i_a} \mathcal{H}_{KI}^{(el)} \tag{112}$$

in Eqs. (95), (99), and (106). These product terms represent a direct coupling between the nuclear kinematic interaction and electronic optical field $V_{IJ}^{(opt)}$. They have arisen from the term $\frac{\partial}{\partial R_{i_a}}(|\Phi_I\rangle \mathcal{H}_{IJ}^{(el)} \langle \Phi_J|)$ of Eq. (93), which serves as a part of the nuclear derivative of the applied electronic potential function. On the other hand, it is interesting to note that the vector potential for the nuclear positions are directly included in the determining equations for electrons through the term $\frac{q_k e}{c}A_k X_{IJ}^k$ in Eq. (20). Here in the electron wavepacket dynamics, we have the direct coupling between the vector potential and nonadiabatic coupling in the Hamiltonian formalism in (R, P)-space. On the other hand, in the presentation in terms of the nuclear variables with $(R, \dot{R}, \ddot{R})$ such a direct coupling is absorbed in the variable $\dot{R}$ by the relation Eq. (109), resulting in Eqs. (110) and (111). Thus the coupling between the vector potential and derivative coupling X_{IJ}^k is realized indirectly only through the variation of the electronic wave function in this representation. In any case, the vector potential for nuclei can affect the propagation of the electronic states through the nonadiabatic coupling.

F. Velocity Form and Length Form?

So far the treatment of classical electromagnetic field in this chapter has been general, and the optical operator appearing in the electronic Hamiltonian, $\hat{\mathcal{H}}^{(el)} = \hat{H}^{(el)} + \hat{V}^{opt}$, has the form of $\hat{V}^{opt} = \frac{1}{2}\sum_j^{el} \left\{ \frac{e}{c}(\hat{p}_j A_j(r_j, t) + A_j(r_j, t)\hat{p}_j) + \frac{e^2}{c^2}A_j(r_j, t)^2 \right\}$. In this section, the Coulomb gauge, $\nabla_{\mathbf{r}} \cdot \mathbf{A} = 0$ is used as usual. In addition, we assume the long wavelength approximation to the vector potential $\mathbf{A}$, which is widely adopted approximation in practice, because the wavelength of light we consider is much longer than the size of the target molecules. This approximation $\mathbf{A}_j(\mathbf{r}_j, t) = \mathbf{A}(t)$ leads to $\hat{V}^{\text{opt}} = \frac{e}{c}A(t) \sum_j^{\text{el}} \hat{p}_j + \frac{1}{2}\frac{e^2}{c^2}A(t)^2 \sum_j^{\text{el}} 1$. We can exclude the second term of $\hat{V}^{\text{opt}}$ because it merely gives rise to the constant phase-shift of the electron wave-packet. Thus our studied optical operator is

$$\hat{V}^{\text{opt}} = \frac{e}{c}A(t) \sum_j^{\text{el}} \hat{\mathbf{p}}_j \tag{113}$$

Using the general form of the vector potential written as the product of the amplitude and the direction vector, $\mathbf{A}(t) = f(t)\hat{\mathbf{u}}(t)$, where $|\hat{\mathbf{u}}| = 1$, the optical matrix becomes $V_{IJ}^{\text{opt}} = \frac{\hbar}{i}\frac{e}{c}f\hat{\mathbf{u}} \cdot \mathbf{D}_{IJ}$. Here, $\mathbf{D}_{IJ} = \left\langle \Phi_I \left| \sum_l^{\text{el}} \frac{\partial}{\partial r_l} \right| \Phi_J \right\rangle$. The electric and magnetic fields are $\mathbf{E} = -\frac{1}{c}\frac{\partial \mathbf{A}}{\partial t} = -\frac{1}{c}\left(\frac{\partial f}{\partial t}\hat{\mathbf{u}} + f\frac{\partial \hat{\mathbf{u}}}{\partial t} \right)$ and $\mathbf{B} = \nabla \times \mathbf{A} = 0$, respectively.

It is well known [53, 79] that making use of the relation $[\hat{\mathbf{r}}, \hat{H}] = i\hbar\mathbf{p}$ along with the eigenfunctions of the electronic Hamiltonian—that is, the adiabatic basis set—the electron momentum matrix can be mathematically transformed from the velocity form to the length form. In this form, the matrix element V_{IJ}^{opt} are evaluated in terms of the dipole moments and the relevant energy gaps [53]. In what follows, we first show that within the fixed nuclei approximation the velocity form is transformed to the length form even for the electronic basis functions other than the adiabatic bases as long as the Coulomb gauge is adopted: Suppose that we have an electronic wavefunction of a molecule, say $\Psi(t)$, whose nuclei are fixed in space. $\Psi(t)$ is therefore determined in a molecular frame. For this function, let us consider the unitary transformation

$$|\Psi^L(t)\rangle = T(t)|\Psi(t)\rangle \tag{114}$$

where

$$T(\mathbf{r}, t) \equiv \exp\left(+\frac{i}{\hbar}\mathbf{r} \cdot \frac{e}{c}\mathbf{A}(t) \right) \tag{115}$$

We designate $|\Psi^L\rangle$ as a total electron state represented in the length form that is transformed from the original $|\Psi\rangle$ in the velocity form representation. Note that this transformation operator $T(\mathbf{r}, t)$ does not depend on $\mathbf{R}$, and we omit even $\mathbf{r}$ below and write it as $T(t)$ for simplicity. First, we treat the Schrödinger equation for one electron without nuclear derivative coupling terms and then consider the case with the coupling terms, which we should examine. The equation of motion for electron is

$$i\hbar \frac{d}{dt}|\Psi(t)\rangle = \left\{\frac{1}{2}\left(\mathbf{p} + \frac{e}{c}\mathbf{A}\right)^2 + V_c(\mathbf{r})\right\}|\Psi(t)\rangle \tag{116}$$

or

$$i\hbar \frac{d}{dt}|\Psi(t)\rangle = \left\{\frac{1}{2}\mathbf{p}^2 + \frac{e}{c}\mathbf{A}\cdot\mathbf{p} + \frac{e^2}{2c^2}\mathbf{A}^2 + V_c(\mathbf{r})\right\}|\Psi(t)\rangle$$

$$\equiv \left\{\frac{1}{2}\mathbf{p}^2 + V_{\mathrm{opt}}(\mathbf{r}, t) + V_c(\mathbf{r})\right\}|\Psi(t)\rangle \tag{117}$$

where we have used $\mathbf{p}\cdot A\Psi(t) = (\mathbf{p}\cdot\mathbf{A})\Psi(t) + \mathbf{A}\cdot(\mathbf{p}\Psi(t)) = \mathbf{A}\cdot\mathbf{p}\Psi(t)$ due to the Coulomb gauge. Substituting $|\Psi(t)\rangle = T^\dagger(t)|\Psi^L(t)\rangle$ into the left-hand side of Eq. (117), we observe

$$i\hbar \frac{d}{dt}|\Psi(t)\rangle = i\hbar\left\{\frac{dT^\dagger(t)}{dt} + T^\dagger(t)\frac{d}{dt}\right\}|\Psi^L(t)\rangle$$

$$= -T^\dagger(t)e\mathbf{r}\cdot\mathbf{E}(\mathbf{r}, t)|\Psi^L(t)\rangle + T^\dagger(t)i\hbar\frac{d}{dt}|\Psi^L(t)\rangle \tag{118}$$

where $\mathbf{E}(t) = -\frac{1}{c}\frac{d}{dt}\mathbf{A}(t)$ is the associated electronic field. On the other hand, the right-hand side (rhs) counterpart is obviously

$$\mathrm{rhs} = \left\{\frac{1}{2}\mathbf{p}^2 + V_{\mathrm{opt}}(\mathbf{r}, t) + V_c(\mathbf{r})\right\}T^\dagger(t)|\Psi(t)^L\rangle \tag{119}$$

Since the commutation relations between $\mathbf{p}$ and $T^\dagger$ is

$$[\mathbf{p}, T^\dagger] = -\frac{e}{c}T^\dagger\mathbf{A} \tag{120}$$

we have

$$[\mathbf{p}^2, T^\dagger] = -2\frac{e}{c}T^\dagger\mathbf{A}\cdot\mathbf{p} + \frac{e^2}{c^2}T^\dagger\mathbf{A}^2 \tag{121}$$

Using this commutation relation, we readily find

$$\left\{\frac{1}{2}\mathbf{p}^2 + V_{\mathrm{opt}}(r, t) + V_c(\mathbf{r})\right\}T^\dagger(t) = T^\dagger(t)\left(\frac{\mathbf{p}^2}{2} + V_c\right) \tag{122}$$

Combining Eqs. (118), (119), and (122) and casting them back into the $\Psi^L(t)$ of Eq. (114), we have the length form accompanied with the dipole moment, that is,

$$ i\hbar\frac{d}{dt}|\Psi^L(t)\rangle = \left\{\frac{1}{2}\mathbf{p}^2-(-e\mathbf{r})\cdot\mathbf{E}(\mathbf{r},t)+V_c(\mathbf{r})\right\}|\Psi^L(t)\rangle \tag{123} $$

Extension to multielectron systems is rather obvious.

The basic situation seems to be the same, even if the Hamiltonian includes the nuclear kinematic couplings. This is because the above proof rests only on the commutation relation for the electronic operators, Eq. (121), and therefore no essential feature should be changed at first sight. For instance, one can rewrite the electronic Schrödinger equation, Eq. (111), as

$$ i\hbar\frac{\partial}{\partial t}|\Psi(t)\rangle = \left(\hat{\mathcal{H}}^{(el)}-i\hbar\dot{\mathbf{R}}\cdot\nabla_\mathbf{R}-\frac{\hbar^2}{2}\nabla_\mathbf{R}^2\right)|\Psi(t)\rangle \tag{124} $$

with

$$ \hat{\mathcal{H}}^{(el)} = \sum_j \frac{\hat{\mathbf{p}}_j^2}{2} + \sum_j\left(\frac{e}{c}\mathbf{A}_j\cdot\mathbf{p}_j+\frac{e^2}{2c^2}\mathbf{A}_j^2\right) + V_c(\mathbf{r};\mathbf{R}) \tag{125} $$

Due to the relation

$$ \left\{\sum_j\frac{\hat{\mathbf{p}}_j^2}{2}+\sum_j\left(\frac{e}{c}\mathbf{A}_j\cdot\mathbf{p}_j+\frac{e^2}{2c^2}\mathbf{A}_j^2\right)+V_c(\mathbf{r};\mathbf{R})-i\hbar\dot{\mathbf{R}}\cdot\nabla_\mathbf{R}-\frac{\hbar^2}{2}\nabla_\mathbf{R}^2\right\}T^\dagger(t) $$

$$ = T^\dagger(t)\left(\sum_j\frac{\hat{\mathbf{p}}_j^2}{2}+V_c-i\hbar\dot{\mathbf{R}}\cdot\nabla_\mathbf{R}-\frac{\hbar^2}{2}\nabla_\mathbf{R}^2\right) \tag{126} $$

which is a direct extension of Eq. (122), it immediately follows that

$$ i\hbar\frac{\partial}{\partial t}|\Psi^L(t)\rangle = \left(\sum_j\frac{\hat{\mathbf{p}}_j^2}{2}-\sum_j(-e\mathbf{r}_j)\cdot\mathbf{E}_j+V_c(\mathbf{r};\mathbf{R})-i\hbar\dot{\mathbf{R}}\cdot\nabla_\mathbf{R}-\frac{\hbar^2}{2}\nabla_\mathbf{R}^2\right)|\Psi^L(t)\rangle \tag{127} $$

Thus, it seems that the velocity form is formally reduced to the length form irrespective of the representation of electronic states, even when the derivative coupling is involved. However, this is not the case: As commented above, the electronic transformation operator T is defined within the molecular frame, which implies that T is implicitly dependent on the nuclear configuration. To be more precise, $T(\mathbf{r}, t)$ should be described as $T(\mathbf{r}, t; \mathbf{R})$, just as the electronic basis functions are so written like $\Phi_I(\mathbf{r}; \mathbf{R})$. Thus T does not commute with the nuclear derivative operators in Eq. (126), and Eq. (127) is at most an approximation, the extent of which is not generally known. We therefore should use the velocity form as long as it is accessible.

G. The Hellmann–Feynman Approximation to the Matrix Element of $\hat{V}^{\mathrm{opt}}(\mathbf{r})$

For the reason stated above, we consistently use the velocity form to represent the electronic matrix element with respect to the vector potential. However, there is a little technical difficulty in the evaluation of $\partial\mathcal{H}_{IJ}^{(\mathrm{el})}/\partial R_{i_a}$ in Eq. (106), and we reduce it to the following simplification. In Eq. (106), consider the term

$$\sum_{IJ} C_I^* \frac{\partial\mathcal{H}_{IJ}^{(\mathrm{el})}}{\partial R_{i_a}} C_J = \sum_{IJ} C_I^* \frac{\partial\left(H_{IJ}^{(\mathrm{el})} + \left\langle \Phi_I \middle| \hat{V}^{\mathrm{opt}} \middle| \Phi_J \right\rangle\right)}{\partial R_{i_a}} C_J \qquad (128)$$

the second term of which is not easy to calculate. On the other hand, Eq. (108) suggests that such a matrix element is not necessary if the basis set at hand were complete. However, it is also known that a direct use of the Hellmann–Feynman force arising from the electronic Hamiltonian is not a good approximation. Therefore we compromise these two factors by approximating Eq. (106) as

$$\left\langle \Phi \middle| \ddot{\mathcal{R}}^{i_a} \middle| \Phi \right\rangle = -\sum_{I,J,K} C_I^* \left(X_{IK}^{i_a} H_{KJ}^{(\mathrm{el})} - H_{IK}^{(\mathrm{el})} X_{KJ}^{i_a} \right) C_J - \sum_{IJ} C_I^* \frac{\partial H_{IJ}^{(\mathrm{el})}}{\partial R_{i_a}} C_J$$
$$- \sum_{I,J} C_I^* \left\langle \Phi_I \middle| \frac{\partial \hat{V}^{\mathrm{opt}}}{\partial R_{i_a}} \middle| \Phi_J \right\rangle C_J + q_a e (\mathbf{E}_a + \dot{\mathbf{R}}_a \times \mathbf{B}_a)_{i_a} \qquad (129)$$

But due to Eq. (103), the term related to $\partial\hat{V}^{\mathrm{opt}}/\partial R_{i_a}$ vanishes, and after all, we have

$$\left\langle \Phi \middle| \ddot{\mathcal{R}}^{i_a} \middle| \Phi \right\rangle = -\sum_{I,J,K} C_I^* \left(X_{IK}^{i_a} H_{KJ}^{(\mathrm{el})} - H_{IK}^{(\mathrm{el})} X_{KJ}^{i_a} \right) C_J - \sum_{IJ} C_I^* \frac{\partial H_{IJ}^{(\mathrm{el})}}{\partial R_{i_a}} C_J$$
$$+ q_a e (\mathbf{E}_a + \dot{\mathbf{R}}_a \times \mathbf{B}_a)_{i_a} \qquad (130)$$

Thus, the force in the mean-field approximation is affected directly by the magnetic Lorentz force and indirectly through the electronic mixing coefficients

$\{C_I\}$. The magnitude of the error in Eq. (130) by totally neglecting $\hat{V}^{\mathrm{opt}}$ is expected to lie around zero anyway. We use this approximation in what follows.

V. NUMERICAL STUDIES OF MOLECULAR SYSTEMS BASED ON THE SEMICLASSICAL EHRENFEST THEORY

Let us explore the nonadiabatic dynamics of molecules placed in an laser field by numerically realizing what is expected to happen. The following numerical examples will give just a first step to start *ab initio* laser chemistry of polyatomic molecules.

A. Computational Details

1. The Vector Potential and Field Intensity

As stated above, we treat the electron-field interaction operator, V^{opt}, in the velocity form, Eq. (113). The vector potential function for a linearly polarized laser pulse employed here is of a Gaussian form:

$$\mathbf{A}(t) = A_s \exp\left(-\left(\frac{t-t_c}{t_w}\right)^2\right)\cos\left(\omega t + \delta\right) \cdot \hat{u} \tag{131}$$

where $\hat{u}$ denotes a constant unit direction vector. Here, t_c, t_w, ω, and δ are the Gaussian central time, time width, frequency, and carrier phase, respectively. This form gives the following electronic field:

$$\mathbf{E}(t) = \frac{A_s\omega}{c}\exp\left(-\left(\frac{t-t_c}{t_w}\right)^2\right)\left(\sin\left(\omega t + \delta\right) + \frac{2(t-t_c)}{\omega t_\omega^2}\cos\left(\omega t + \delta\right)\right) \cdot \hat{u} \tag{132}$$

According to the usual convention, we define the electric field strength E_s as

$$E_s \equiv \frac{\omega}{c}A_s \tag{133}$$

and it is also the standard practice to define the field intensity as

$$I \equiv \frac{\epsilon_0\omega^2 A_s^2}{2c} = \frac{c\epsilon_0 E_s^2}{2} \tag{134}$$

Here, ϵ_0 is the dielectric constant in the vacuum. Our practice in this article is to specify E_s first to label the field strength, and then convert it to A_s in Eq. (133). Then we actually use Eq. (131) to carry out the computations.

For a linear combination of two fields,

$$\mathbf{A}(t) = A_{s1}\exp\left(-\left(\frac{t-t_{c1}}{t_{w1}}\right)^2\right)\cos\left(\omega_1(t-t_{c1})+\delta_1\right)\cdot\hat{u}_1$$
$$+A_{s2}\exp\left(-\left(\frac{t-t_{c2}}{t_{w2}}\right)^2\right)\sin\left(\omega_2(t-t_{c2})+\delta_2\right)\cdot\hat{u}_2$$

(135)

or more, the intensity of the entire synthesized field is referred to as the set of the individual intensities defined as above. In what follows, we have set all the phases $\delta = 0$, $\delta_1 = 0$ and $\delta_2 = 0$ just for simplicity. (The variation of the phases is sometimes critical in dynamics [80].)

2. Molecular Systems Studied

To examine the effects of rather intense laser fields on the various dynamics of molecules such as electronic state population, electronic current within a molecule, geometrical change, and chemical reaction, we choose the following four species; methane, methyl alcohol, diborane, and $(LiH_2)^+$. Methane, methyl alcohol, and diborane are placed in the coordinate systems of Figs. 1, 2 and 3, respectively. The potential surfaces of $LiH + H^+/LiH^+ + H$ and $Na^+ + Cl^-/Na + Cl$ will be presented later.

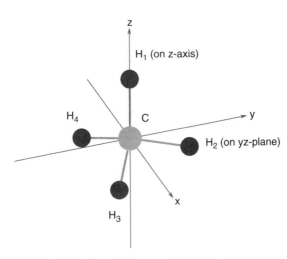

Figure 1.   Methane (CH$_4$) in the coordinates.

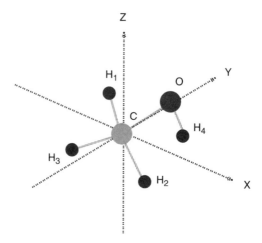

Figure 2. Methyl alcohol (CH$_3$OH) in the coordinates.

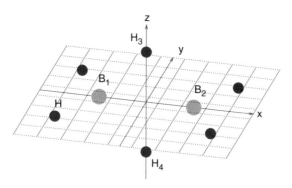

Figure 3. Diborane (H$_2$BH$_2$BH$_2$) in the coordinates.

3. Field Parameters and Conversion Table

The parameters used for lasers are listed in Table 1 along with the conversion table from the absolute units to the conventional ones. The effective electric field strength E_s and field intensity I are defined in Eqs. (133) and (134), respectively. In this section, time and bond distance are measured in units of femtosecond and bohr (0.529 Å), respectively.

4. Level of Electronic State Calculations

The level of *ab initio* calculations for molecular orbitals and the configuration functions (CSF) are summarized for the above molecular systems. (1) Methane:

TABLE I
Conversion Table for Field Parameters

ω (a.u.)	$\hbar\omega$ (eV)	E_s (a.u.)	I (W/cm^2)
0.057	1.55	0.02	1.40×10^{13}
0.228	6.20	0.03	3.15×10^{13}
0.300	8.16	0.04	5.60×10^{13}
0.516	14.04	0.06	1.26×10^{14}
		0.08	2.24×10^{14}
		0.10	3.50×10^{14}

RHF/CIS/6-31G(d,p), 35 RHF molecular orbitals (MOs), 121 configurational state functions (CSFs), core C $1s$ orbital. (2) Methyl alcohol: RHF/CIS/6-31G(d), 38 MOs, 204 CSFs, core C, O $1s$. (3) Diborane: RHF/CIS/6-31G(d), 42 MOs, 205 CSFs, core B $1s$. (4) (LiH$_2$)$^+$: RHF/CISD/6-31G, 13 MOs, 78 CSFs, core Li $1s$. (5) NaCl: RHF/CISD/6-31G, 26 MOs, 91 CSFs, core orbitals up to HOMO-1.

In solving the coupled dynamics including the external field, we employed (a) the sixth-order Gear predictor–corrector algorithm [81] for nuclear propagation with the time step $dt_{nuc} = 0.025$ fs and (b) the Chebyshev polynomial expansion method [82] for electron time propagator with $dt_{elec} = dt_{nuc}/100$. At each time of dt_{nuc}, the electronic Hamiltonian matrix was constructed at the renewed nuclear positions. The optical terms are updated at each time step for electronic propagation, dt_{elec}. In other words, for a very short time interval of dt_{elec} to propagate the electron wavepacket, the electronic Hamiltonian matrix arising from the pure electronic Hamiltonian was frozen in time.

B. Response of Bond Length and Bond Order of Methane to Laser Frequency

First we study the basic response of the molecules to laser frequency. Our naive question is in which laser frequency ω, high or low, the nonadiabatic coupling X_{IJ} will take a large responsibility for the change of molecular geometry. Figure 4 shows the time-dependent change of the bond length of C–H$_1$ in the ground state of methane. For Fig. 4a (the first and third) and Fig. 4b (the second and fourth), lasers of $\omega = 0.057$ and $\omega = 0.516$ are applied. The frequency $\omega = 0.516$ corresponds to the resonance value of vertical transition from the ground to first electronic excited states at the stable geometry, while $\omega = 0.057$ is in the infrared (IR) range. The upper two panels exhibit the dynamics under the electronic field intensity $E_s = 0.02$, while the lower two panels are for $E_s = 0.06$. In the individual panels, two comparative calculations are shown; one with the nonadiabatic coupling X_{IJ} (solid curve) and the other without it (dotted curve).

It is obvious and quite understandable that the higher-energy laser ($\omega = 0.516$) induces much larger geometrical change. On the other hand, it is clearly illustrated

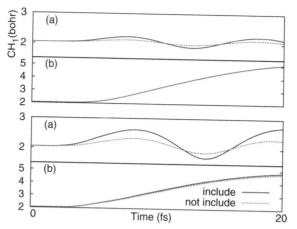

Figure 4.   Effect of the nonadiabatic coupling X_{IJ}^k on the time-dependent distance of a C–H bond of methane. The upper pair and lower pair of panels represent the case of $E_s = 0.02$ and 0.06, respectively. Panels (a) and (b) are for $\omega = 0.057$ and 0.516, respectively. In each panel, the solid (dotted) curves show the result in the presence (absence) of the laser field.

that the effect of the nonadiabatic coupling X_{IJ} manifests itself much more clearly in the lower-frequency case (Fig. 4a). The effect of the indirect coupling between electronic and derivative coupling interactions via Eq. (108) and Eq. (111) emerges in the nuclear dynamics. It seems natural that the frequency in the IR range should pick more direct influence from the nonadiabatic interaction to the molecular geometry. On the other hand, the effect of X_{IJ} seems to be much more limited for the higher-frequency laser.

Let us next look at the dynamics of bond order of methane ground state in Fig. 5. Again, Fig. 5a (the first and third) and Fig. 5b (the second and fourth) include the bond order dynamics for the lasers of $\omega = 0.057$ and $\omega = 0.516$. The upper panels are for the dynamics under the electronic field intensity $E_s = 0.02$, while the lower two panels are for $E_s = 0.06$. As far as the these field intensities are concerned, the effect of X_{IJ} on the bond order turns out to be very small except in the case of $\omega = 0.057$ and $E_s = 0.06$ (the third panel).

In discussion about the effect of the nonadiabatic coupling though, we should be very careful about what representation, or which kind of basis functions, is taken for propagation of the electronic wavepacket. In this regard, we re-emphasize that we consistently use the CSF (configuration state function) basis and that this basis set provides a very good (not perfect though) diabatic representation. This fact has been clearly established before [37]. Therefore, if one uses adiabatic electronic functions to expand the electron wavepackets, the effect of the nonadiabatic coupling is totally different and expected to be must larger.

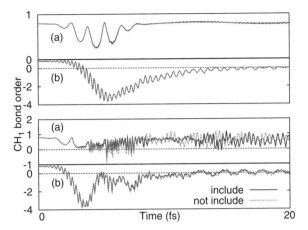

Figure 5.   Effect of the nonadiabatic coupling X_{IJ}^k on the time-dependent bond order of a CH bond of methane. The configuration of the panels is the same as that of Fig. 4

In view of the fact that the effect of the nonadiabatic coupling manifests itself more significantly in the IR regime, we mainly choose this frequency in what follows.

C.   Effects of Laser Intensity in the Presence of the Nonadiabatic Interaction

The above observation, that is, the small effect of the nonadiabatic coupling in the CSF basis, is not necessarily the case in the high intensity regime. This is simply because the molecular response is not linear in the intensity. We resume the calculation of methane ground state under the laser frequency of $\omega = 0.057$, varying the intensity as $E_s = 0.02, 0.04, 0.06$, and 0.08. Figures 6, 7, and 8 record the dynamics of molecular geometry, the electronic-state population for the ground state (unity before shining the laser), and the total energy of the molecule, respectively. In all the figures, the upper panels take account of the nonadiabatic coupling X_{IJ} explicitly, while the lower panels do not.

A remarkable dependence of the X_{IJ} effect on the laser intensity is observed in the figures. We observe a large transition of the molecular response to the laser in between $E_s = 0.04$ and $E_s = 0.06$. The effect becomes more prominent as time passes longer. Therefore one needs to take an explicit account of the nonadiabatic coupling in the study of intense laser chemistry even for ground-state dynamics, particularly in the calculation of long time dynamics of molecular geometry.

D.   Question to the Validity of Fixed Nuclei Approximation
in Intense Laser Field

In the study of electron wavepacket dynamics under intense laser field, it is very attractive to assume that the pulse laser of a width shorter than, say, 10 fs is so much

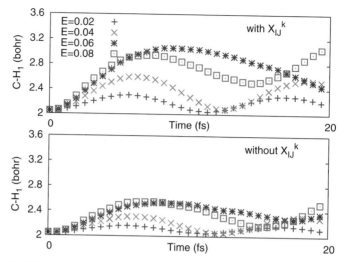

Figure 6. Time-dependent C–H distance of methane along a field polarization parallel to this CH with X_{IJ}^k (**upper panel**) and without X_{IJ}^k (**lower panel**). Continuum IR ($\omega = 0.057$) laser field with the field strength of $E_s = 0.02, 0.04, 0.06,$ and 0.08 are applied.

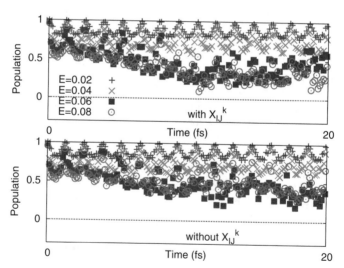

Figure 7. Variation of the population of the electronic ground state of methane with X_{IJ}^k (**upper panel**) and without X_{IJ}^k (**lower panel**). Continuum IR ($\omega = 0.057$) laser field with the field strength of $E_s = 0.02, 0.04, 0.06,$ and 0.08 are applied.

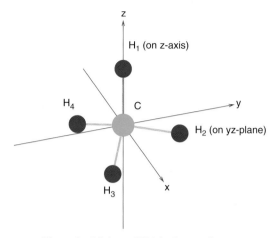

Figure 1. Methane (CH₄) in the coordinates.

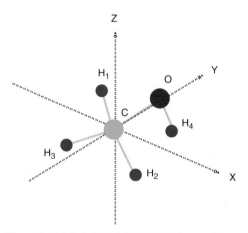

Figure 2. Methyl alcohol (CH₃OH) in the coordinates.

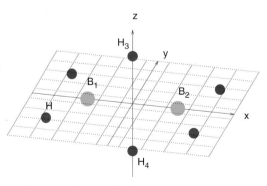

Figure 3. Diborane (H₂BH₂BH₂) in the coordinates.

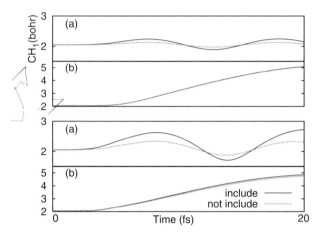

Figure 4. Effect of the nonadiabatic coupling X_{IJ}^k on the time-dependent distance of a C–H bond of methane. The upper pair and lower pair of panels represent the case of $E_s = 0.02$ and 0.06, respectively. Panels (**a**) and (**b**) are for $\omega = 0.057$ and 0.516, respectively. In each panel, the solid (dotted) curves show the result in the presence (absence) of the laser field.

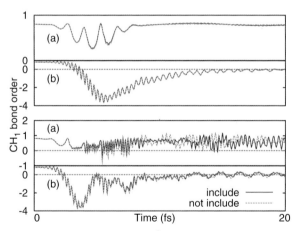

Figure 5. Effect of the nonadiabatic coupling X_{IJ}^k on the time-dependent bond order of a CH bond of methane. The configuration of the panels is the same as that of Fig. 4

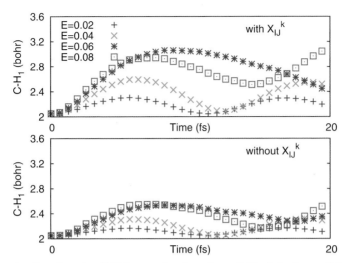

Figure 6. Time-dependent C–H distance of methane along a field polarization parallel to this CH with X_{IJ}^k (**upper panel**) and without X_{IJ}^k (**lower panel**). Continuum IR ($\omega = 0.057$) laser field with the field strength of $E_s = 0.02, 0.04, 0.06,$ and 0.08 are applied.

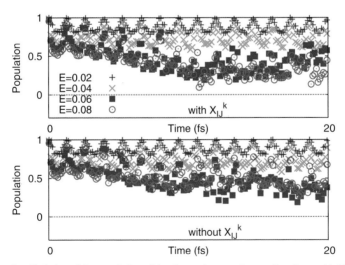

Figure 7. Variation of the population of the electronic ground state of methane with X_{IJ}^k (**upper panel**) and without X_{IJ}^k (**lower panel**). Continuum IR ($\omega = 0.057$) laser field with the field strength of $E_s = 0.02, 0.04, 0.06,$ and 0.08 are applied.

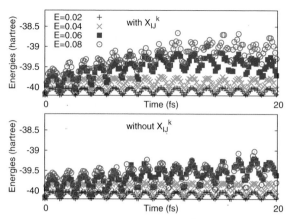

Figure 8. Variation of the total energy of methane with X_{IJ}^k (**upper panel**) and without X_{IJ}^k (**lower panel**). Continuum IR ($\omega = 0.057$) laser field with the field strength of $E_s = 0.02, 0.04, 0.06,$ and 0.08 are applied.

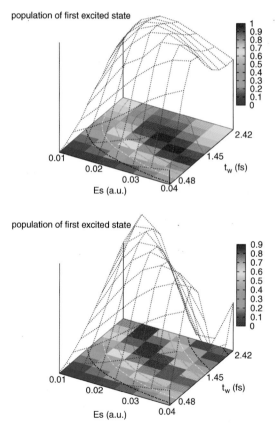

Figure 9. Diagram of the populations of the first electronic excited state of methane as a function of the field intensity and duration of application of laser with $\omega = 0.516$. (*See text for full caption.*)

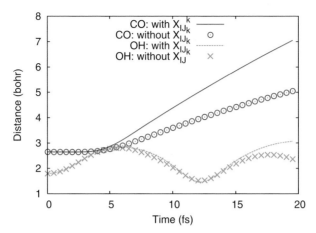

Figure 10. CO and OH distances of methyl alcohol, initially prepared in the first electroctronic excited state, under the intense IR laser, whose electronic field is polarized in the direction of CO. The effect of the nonadiabatic coupling is clarified by comparison.

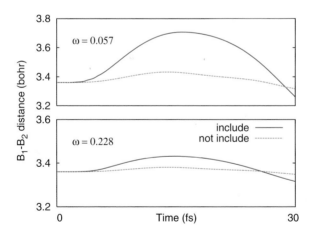

Figure 11. Vibrational excitation of diborane in the laser fields is assisted by the nonadiabatic coupling. The B–B bond distance driven by the z–x circular polarized pulse of a few cycles. The frequencies are $\omega = 0.057$ and 0.228 for the upper and lower panels, respectively. The duration time is approximately 5 fs. The strength of electronic field for each direction is 0.03.

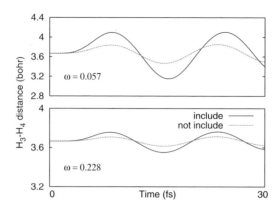

Figure 12. Distance of H_3–H_4 in diborane driven by the z–x circular polarized pulse of a few cycles as in Fig. 11

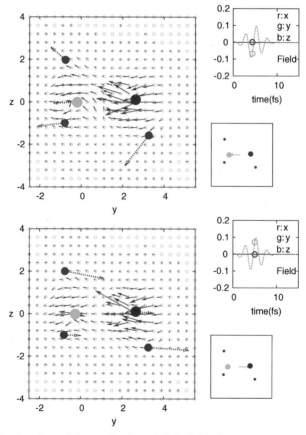

Figure 13. Snapshots of electron flux in methyl alcohol in the ground state, taken at $t = 4.375$ (**upper panels**) and 5.000 fs (**lower panels**). The IR laser is shined with a polarization vector parallel to the CO bond.

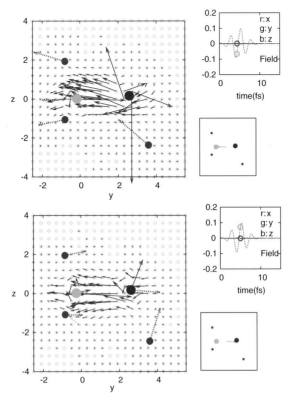

Figure 14. Electron flux in methyl alcohol initially prepared in the first electronic excited state. Snapshots are taken at $t = 4.375$ (**upper panels**) and 5.000 fs (**lower panels**). The IR laser is shined with a polarization vector parallel to the CO bond.

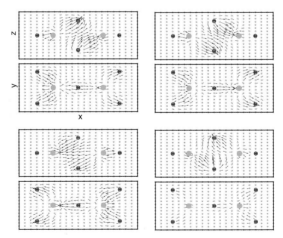

Figure 15. Bond order electron flux of diborane taken at $t = 4.25$, 5.00, 5.75, and 6.50 fs in the upper left, upper right, lower left, and lower right panels, respectively. $\omega = 0.057$ and the duration time is approximately 5 fs. $E_s = 0.03$. The circular polarization is applied on the x–z plane.

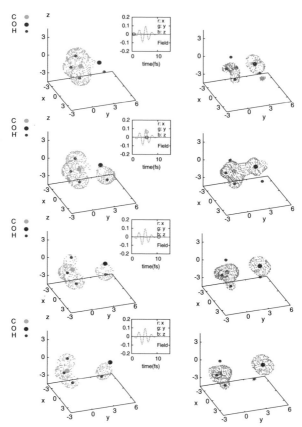

Figure 16. Bond order density (**left column**) and unpaired electon density (**right column**) along the dynamics under the intense IR laser along CO direction for methyl alcohol, which was initially prepared in the first electronic excited state. The CO bond is now broken. Values equal to 0.02 of the densities are plotted.

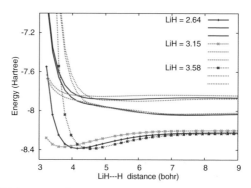

Figure 17. $(LiH_2)^+$ system, adiabatic potential energy curves for the distance between the center of mass of LiH and H in a linear geometry. Three selected cases are plots for LiH length = 2.64, 3.15, and 3.58 bohrs. The ground state is correlated asymptotically to $(Li-H_{(1)})^+ H_{(2)}$, while the first excited state is to $(Li-H_{(1)})H_{(2)}^+$.

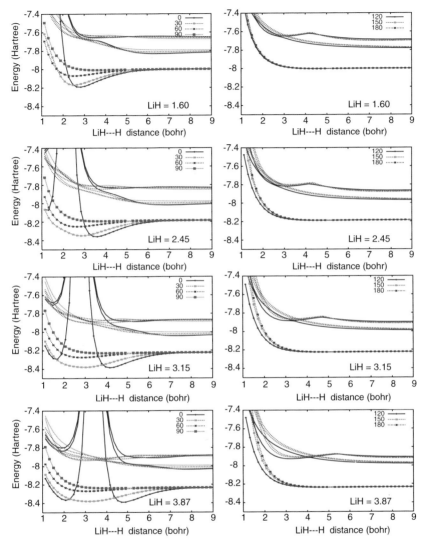

Figure 18. $(LiH_2)^+$ system; adiabatic potential energy curves for the distance between the center of mass of LiH and H with the inserting angle with respect to the LiH and colliding H atom for the LiH length = 1.60, 2.45, 3.15, and 3.87 bohrs.

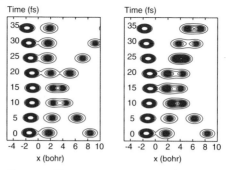

Figure 19. Effect of laser on collinear collision of LiH + H$^+$. **(Left panel)** Electronically elastic scattering of LiH + H$^+$ in the first excited state without laser. **(Right panel)** Abstraction reaction under the laser field of $E_S = 0.02$ a.u., $\omega = 0.3$ a.u., $t_c = 7.26$ fs, $t_w = 2.42$ fs. The electric field is linearly polarlized along the collision coordinate.

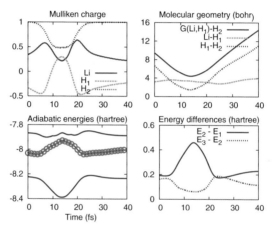

Figure 20. Variation of some relevant quantities during the collinear collision of LiH + H$^+$ in the first excited state without external field.

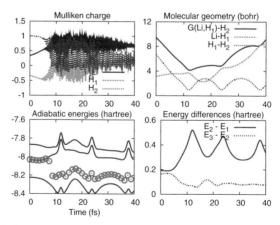

Figure 21. Variation of some relevant quantities during the collinear collision of LiH + H$^+$ in the first excited state under the laser field of $E_S = 0.02$, $\omega = 0.3$, $t_c = 7.26$ fs, $t_w = 2.42$ fs.

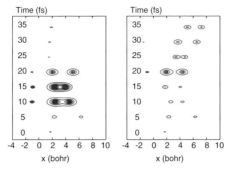

Figure 22. Unpaired electron density to characterize the difference in the collinear collision of LiH + H$^+$ in Fig. 19. (*See text for full caption.*)

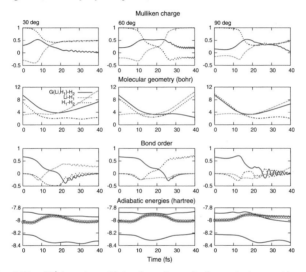

Figure 23. LiH + H$^+$ inserting collision dynamics on the first excited state without external field. The circles in the panel for adiabatic energy denote mean electronic energy.

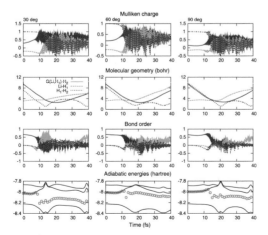

Figure 24. LiH + H$^+$ inserting collision dynamics on the first excited state under the laser field. (*See text for full caption.*)

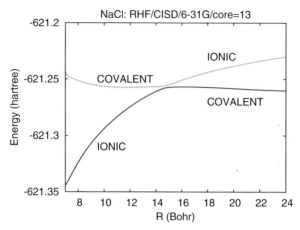

Figure 25. Potential energy curves for the system of Na + Cl around the avoided crossing.

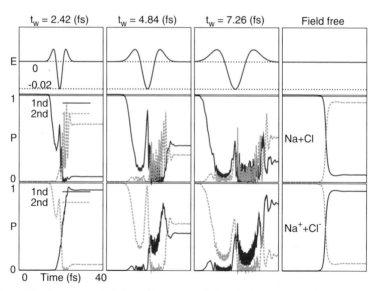

Figure 26. Laser control of electronic state populations for the collisions of Na + Cl (**middle row**) and Na$^+$ + Cl$^-$ (**bottom row**). The electronic populations after crossing the avoided crossing are drawn. The electric fields made by the laser with differnt width are shown in the top row.

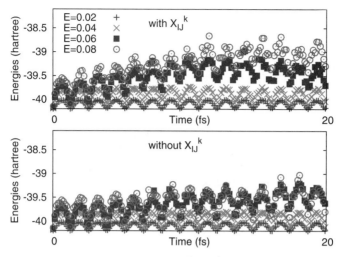

Figure 8. Variation of the total energy of methane with X_{IJ}^k (**upper panel**) and without X_{IJ}^k (**lower panel**). Continuum IR ($\omega = 0.057$) laser field with the field strength of $E_s = 0.02, 0.04, 0.06,$ and 0.08 are applied.

shorter than the time scale of the nuclear dynamics that the nuclei may be fixed in space to see the behavior of the electronic transition. Here we examine the validity of this charming assumption. Taking methane as an example, we have calculated the electronic population to be observed in the first excited state by pumping up from the ground state as a function of the field intensity and pulse width (corresponding to the duration time of shining laser) at a fixed frequency $\omega = 0.516$. Figure 9 shows such a phase diagram; in the upper panel the nuclei can move according to the semiclassical Ehrenfest dynamics under the field, while in the lower panel they are fixed in space at the bottom of the ground state potential energy surface. Obviously they are significantly different. It is seen that in the fixed nuclei approximation, the weaker intensity with longer pulse is favorable for the first excited state to be populated most. On the other hand, in case of dynamically allowed nuclear motion, the stronger and shorter pulse is favorable. Of course, it depends on the physical target one looks at whether this difference is significant or not. In the study of laser control of the population dynamics, at least, the fixed nuclei approximation in an intense laser field is not likely to be a valid premise.

E.   Enhancement of Geometrical Change by the Nonadiabatic Coupling

It is well known that the nonadiabatic coupling elements play the very essential role in the nonadiabatic transition-like electron transfer reaction such as

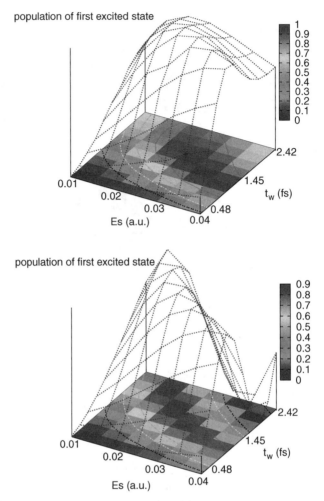

Figure 9.   Diagram of the populations of the first electronic excited state of methane as a function of the field intensity and duration of application of laser with $\omega = 0.516$. While the molecule in the upper figure are allowed to move in space, the molecule in the lower figure is fixed in space. There is no way for the nonadaibatic coupling element to play a role in the fixed nuclear approximation.

$Na + Cl \rightarrow Na^+ + Cl^-$ (we will discuss the control of this reaction later in this section). For other dynamics on a singly isolated potential energy surface, the nonadiabatic coupling is negligibly small and therefore the Born–Oppenheimer is undoubtedly valid. But, this is never the case when a molecular is place in an intense laser field. Below we explore how the nonadiabatic coupling enhances or suppresses the effect of laser.

1. Methyl Alcohol Bond-Distance Dynamics

Shining the laser of $\omega = 0.057$ and $E_s = 0.1$ on methyl alcohol, which was in the first electronic state, we trace the bond length of CO and OH. The pulse width $(2t_w)$ is 4.84 fs, and the optical field is polarized parallel to the CO bond and this bond is cleaved. In Fig. 10, we observe a large effect the nonadiabatic coupling elements. By the presence of X_{IJ}, the OH group is more efficiently driven to recede from the rest of molecule with a higher speed.

2. Geometry of Diborane

We next examine the dynamics of the geometrical change of the ground state of diborane. Diborane is known to have electron-deficient chemical bonds, and thereby it is expected to have an interesting laser response not only in the electronic structure but also in vibronic coupling. Figures 11 and 12 track the bond distances of B_1–B_2 and H_3–H_4, respectively. A laser of circular polarization on the z–x plane with an electric intensity $Es = 0.03$ is applied. In both figures, the frequencies applied are $\omega = 0.057$ and $\omega = 0.228$ for the upper and lower panels, respectively, and the pulse width is 4.84 fs. The calculations including the nonadiabatic coupling elements and excluding them are compared on the same basis. Again, the geometrical change, vibrational excitation, is much more enhanced by the presence of X_{IJ}. Since the nuclei gain more momenta by the laser of IR frequency $\omega = 0.057$ than that of $\omega = 0.228$, the effect of X_{IJ} is larger for the former case. In particular, quite a large amplitude variation in the distance of H_3–H_4 under the laser

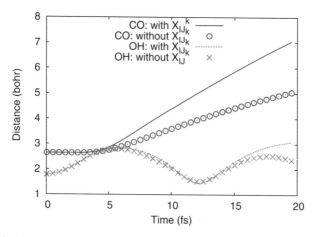

Figure 10.   CO and OH distances of methyl alcohol, initially prepared in the first electroctronic excited state, under the intense IR laser, whose electronic field is polarized in the direction of CO. The effect of the nonadiabatic coupling is clarified by comparison.

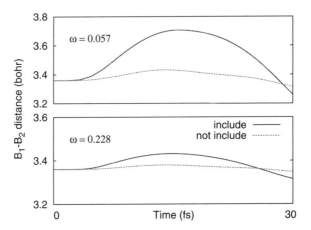

Figure 11. Vibrational excitation of diborane in the laser fields is assisted by the nonadiabatic coupling. The B–B bond distance driven by the $z-x$ circular polarized pulse of a few cycles. The frequencies are $\omega = 0.057$ and 0.228 for the upper and lower panels, respectively. The duration time is approximately 5 fs. The strength of electronic field for each direction is 0.03.

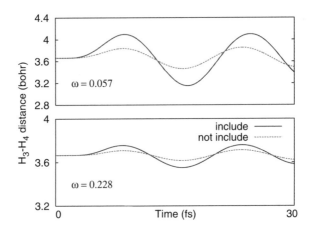

Figure 12. Distance of H_3-H_4 in diborane driven by the $z-x$ circular polarized pulse of a few cycles as in Fig. 11

of $\omega = 0.057$ is noticed. Therefore, the nonadiabatic coupling is vital for the correct estimate of molecular dynamics even in the study of ground state when applied an intense laser.

As far as we have experienced, the presence of the nonadiabatic coupling seems to assist the IR laser to enhance the geometrical change of molecules.

F. Electron Flux Induced by Laser

In our future design of chemical reaction control by reshaping electronic states using intense lasers, the knowledge of (nonadiabatic) electronic response to the fields should become indispensable. Here we show some examples of such response in terms of the electron flux within a molecule.

Suppose we have an electronic wavepacket $\Phi(\mathbf{r}; \mathbf{R}(t))$. The standard flux is defined as follows [83]. With the N-particle differential operator $\nabla_N = \sum_i^N \nabla_i$, the flux in the $3N$-dimensional space is represented as

$$\vec{j}_N(\mathbf{r}_1, \mathbf{r}_2, \ldots, \mathbf{r}_N) = \frac{\hbar}{2i}(\Phi(\mathbf{r}; \mathbf{R}(t))^* \nabla_N \Phi(\mathbf{r}; \mathbf{R}(t)) - \Phi(\mathbf{r}; \mathbf{R}(t))^* \nabla_N \Phi(\mathbf{r}; \mathbf{R}(t)))$$

(136)

at each nuclear position $\mathbf{R}(t)$. This flux is reduced to the one-particle counterpart

$$\vec{j}(\mathbf{r}_1) = \frac{\hbar}{2i}[\nabla_1 \gamma(\mathbf{r}'_1, \mathbf{r}_1; t) - \nabla_1 \gamma(\mathbf{r}_1, \mathbf{r}'_1; t)]|_{\mathbf{r}'=\mathbf{r}}$$

(137)

where $\gamma(\mathbf{r}_1, \mathbf{r}'_1; t)$ is the first-order off-diagonal spin-free density matrix, and ∇_1 is to be operated on the $\mathbf{r}_1$ coordinates only, and then $\mathbf{r}'_1$ is replaced with $\mathbf{r}_1$. Since $\vec{j}(\mathbf{r}_1)$ is identically zero, when a wavefunction is used, it is real-valued as most of the adiabatic electronic wavefunctions are. On the other hand, there are so many studies on electron currents induced by an external field [84, 85, 86]. To track electron current induced by chemical reactions in vacuum, one should find a way to correlate an electron wavepacket with the absolute time by synchronizing nuclear motions. Indeed, Barth et al. have recently extracted electronic and nuclear simultaneous fluxes using the full quantum non-Born–Oppenheimer calculation of H_2^+ molecule [87]. Okuyama and Takatsuka have been systematically investigating electron current induced by chemical reactions, in which the semiclassical Ehrenfest electronic wavepackets, synchronizing nuclear motions through the nuclear kinematic coupling, are used [88] (incidentally, refer to this paper for an extension of the notion of flux so as to retrieve the information of electron current from adiabatic electronic wavefunctions). The electron currents shown below are induced by laser and by nuclear coupling.

1. Induced Electron Flux in Methyl Alcohol in the Ground and First Excited States

With a laser of $\omega = 0.057$ and $E_s = 0.1$ having a pulse width of $5\,\mathrm{fs}$ linearly polarized parallel to the CO bond, we induce electron currents in the ground and first excited states of methyl alcohol, which are shown in Figs. 13 and 14, respectively. The fluxes on the y–z plane, a plane including COH_4, is shown for both cases. In both figures, the upper and lower panels display the flux at time

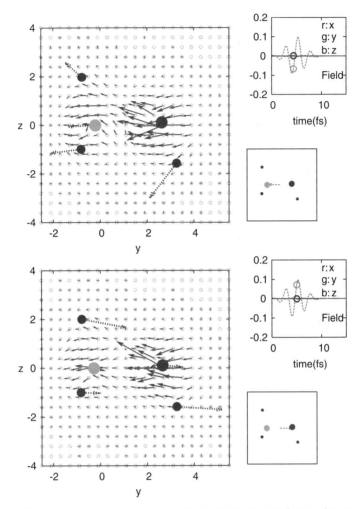

Figure 13. Snapshots of electron flux in methyl alcohol in the ground state, taken at t = 4.375 (**upper panels**) and 5.000 fs (**lower panels**). The IR laser is shined with a polarization vector parallel to the CO bond.

$t = 4.375$ fs and $t = 5.0$ fs. The directions of electron flux are the same in these snapshots both for the two cases, while the direction of the electric field turns the other way around. At $t = 4.375$ fs, the nuclei are basically pulled to the left, while at $t = 5.0$ fs the forces working on them are directed mostly to the right, which are drawn with broken lines on the individual nuclei. The electron flux vectors are drawn on the mesh points. The current nearby the oxygen atom is significantly large, which should represent a current within the atom.

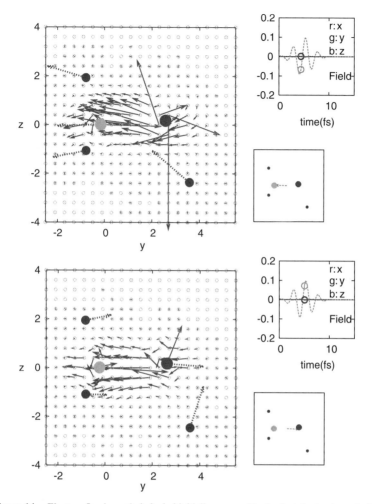

Figure 14. Electron flux in methyl alcohol initially prepared in the first electronic excited state. Snapshots are taken at $t = 4.375$ (**upper panels**) and 5.000 fs (**lower panels**). The IR laser is shined with a polarization vector parallel to the CO bond.

Two phenomena are immediately observed. In comparing the upper and lower columns in each state (in each figure), we notice that the electronic current does not respond to the directional change of the electric field very quickly. Rather, the electrons behave as though they have large inertia. For the first excited state (Fig. 14). the response is a little faster, presumably reflecting its more flexible electronic state. In both cases, it seems that the field is not intense enough and/or the frequency is not high enough to dominate the electronic motion against their intrinsic dynamics within the molecule. The other observation is that the electron current in the ground state is roughly grouped in the areas of the functional group CH_3 and OH, while the electron current induce in the first excited state is more global, that is, it flows between OH to CH_3 areas. The notion of functional group is robust in this sense, too.

2. Induced Electron Flux in Diborane

We next explore induced electron flux in the electron-deficient chemical bonding of diborane. As in Figs. 11 and 12, we shine a laser of circular polarization on the z–x plane with $E_s = 0.03$, $\omega = 0.057$, and the pulse width 5 fs. Plotted in Fig. 15 are the bond-order type electron flux, which is defined as follows. The flux of Eq. (137) can be expanded in the atomic orbitals as

$$\vec{j}(\mathbf{r}) = \frac{\hbar}{2i} \sum_{aA} \sum_{bB} \rho_{aA,bB} [\chi_{aA}(\mathbf{r}) \nabla \chi_{bB}(\mathbf{r}) - \chi_{bB}(\mathbf{r}) \nabla \chi_{aA}(\mathbf{r})] \qquad (138)$$

where $\chi_{aA}(\mathbf{r})$ is an atomic orbital a on an atom A. However, it quite often happens that the electron flux within an atom not only is irrelevant to bond rearrangement but also makes the graphic representation unclear. Indeed we have seen such an example in Figs. 13 and 14. Therefore, to concentrate on the electron flux that is

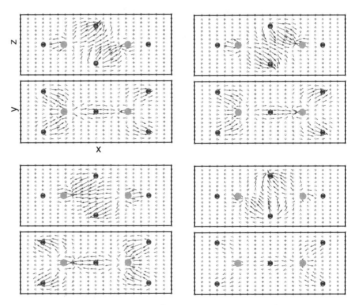

Figure 15. Bond order electron flux of diborane taken at $t = 4.25$, 5.00, 5.75, and 6.50 fs in the upper left, upper right, lower left, and lower right panels, respectively. $\omega = 0.057$ and the duration time is approximately 5 fs. $E_s = 0.03$. The circular polarization is applied on the x–z plane.

relevant to chemical bonding, we display, instead of $\vec{j}(\mathbf{r})$, the following quantity:

$$\vec{j}_{bond}(\mathbf{r}) = \frac{\hbar}{2i} \sum_{a}^{on\ A} \sum_{b}^{on\ B(\neq A)} \rho_{aA,bB}[\chi_{aA}(\mathbf{r})\nabla\chi_{bB}(\mathbf{r}) - \chi_{bB}(\mathbf{r})\nabla\chi_{aA}(\mathbf{r})] \qquad (139)$$

which we call the bond-order flux. In Fig. 15, the bond-order flux at $t = 4.25$, 5.00, 5.75, and 6.50 fs is presented in the upper left, upper right, lower left, and lower right panels, respectively. Each panel is composed of two representations: The upper is the side view for z–x plane including B–B bond, while the lower is the top looking at y–x plane including B–B bond. In response to the circular polarization of the laser applied, the electron current induced is seen to follow the circular change of the vector potential very quickly. This makes a remarkable contrast to the slow response in methyl alcohol observed as above. The circular polarization for $\omega = 0.057$ gives a period of circulation about 2.66 fs, which is consistent with the period actually observed. Incidentally, it is interesting to confirm that the current in the area of BH_2 outside the B–B bond resonates the flux within the B–B bond.

A question remains as to the mechanism of inducing the flux in diborane. There are two driving forces to induce the electron current: One is the electric field provided by the laser, and the other is the geometrical dislocation of the nuclei almost in resonance to the IR laser. In view of the insensitive response of the electron current to the fields in methyl alcohol, the latter mechanism is likely to be predominant. More of this aspect will be discussed elsewhere.

G.   Dissociation and Collision in Laser Field

As a next illustrative example of nonadiabatic electron wavepacket dynamics in laser fields, we explore a couple of elementally chemical reactions. These set a preliminary foundation of the control of chemical reactions with the optical fields.

1.   Photodissociation of CH_3OH

We begin with a simple (single-photon) photodissociation of methyl alcohol via the first excited state. As a simple photodissociation, the reaction proceeds as

$$CH_3OH + h\nu \rightarrow CH_3O + H.$$

But, as already suggested in Fig. 10, upon shining the pulse laser, it changes to

$$CH_3OH + h\nu + IR-laser \rightarrow CH_3 + OH$$

Figure 16 (left column) tracks the bond order density, which is defined as

$$\gamma_{\text{bond}}(\mathbf{r}) = \sum_{a}^{\text{on A}} \sum_{b}^{\text{on B}(\neq A)} \rho_{aA,bB}[\chi_{aA}(\mathbf{r})\chi_{bB}(\mathbf{r})] \tag{140}$$

It is clear from this figure that the laser has canceled the bond between C and O and instead creates bonding between O and H. The bond of OH thus receding from the methyl group is vibrationally excited.

In the right column of this figure, we have plotted what we call unpaired electron density, whose definition is [89]

$$D(\mathbf{r}) = 2\gamma(\mathbf{r}) - \int d\mathbf{r}' \gamma(\mathbf{r}, \mathbf{r}')\gamma(\mathbf{r}', \mathbf{r}) \tag{141}$$

where $\gamma(\mathbf{r}, \mathbf{r}')$ is the spin-free off-diagonal first-order density matrix and $\gamma(\mathbf{r}) = \gamma(\mathbf{r}, \mathbf{r})$. $D(\mathbf{r})$ is an indicator of the spatial distribution of the electrons that do not make pairing by α-spin and β-spin in natural orbitals. It works also as an indicator of the extent of electron correlation. For instance, the Hartree–Fock single determinant singlet state identically gives $D(\mathbf{r}) = 0$. In Fig. 16, it is understood that the laser has made a diradical state, the radicals of which are localized in methyl group and OH group separately, which has driven the dissociation CH$_3$ + OH. In this way, the laser has rearranged the electronic nature to change the course of chemical reaction.

2. Collision of LiH + H$^+$

We next study a very basic and simple molecular collision of LiH + H$^+$ in laser field. The potential surfaces for (LiH$_2$)$^+$ system are drawn in Figs. 17 and 18.

Collinear Collision of LiH + H$^+$. The collinear collision under the field free condition undergoes simply an (electronically) elastic scattering, that is,

$$\text{LiH} + \text{H}^+ \rightarrow \text{LiH} + \text{H}^+ \tag{142}$$

This is actually the first excited state in the asymptotic region. The relative kinetic energy is 0.1 hartree throughout all the following collisions. The electron density map in the course of the reaction is shown in the left column of Fig. 19. (Note that time proceeds from bottom to top.) On this system, we shine the laser of $E_s = 0.02$, $\omega = 0.3$, $t_c = 7.26$ fs, $t_w = 2.42$ fs, setting the polarization vector on the molecular axis. The timing is adjusted so that the laser peak becomes maximum around the time when H$^+$ begins to climb the gradual potential slope. The reaction has

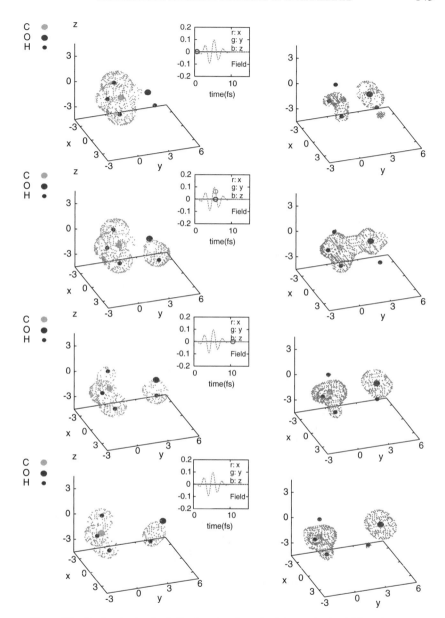

Figure 16. Bond order density (**left column**) and unpaired electon density (**right column**) along the dynamics under the intense IR laser along CO direction for methyl alcohol, which was initially prepared in the first electronic excited state. The CO bond is now broken. Values equal to 0.02 of the densities are plotted.

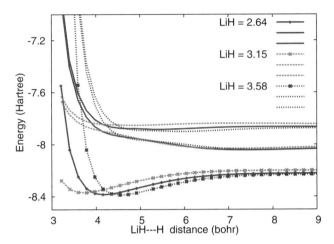

Figure 17. $(LiH_2)^+$ system, adiabatic potential energy curves for the distance between the center of mass of LiH and H in a linear geometry. Three selected cases are plots for LiH length = 2.64, 3.15, and 3.58 bohrs. The ground state is correlated asymptotically to $(Li-H_{(1)})^+ - H_{(2)}$, while the first excited state is to $(Li-H_{(1)})-H_{(2)}^+$.

changed to

$$LiH + H^+ \rightarrow \begin{cases} \rightarrow LiH + H^+ \\ \rightarrow Li + H_2^+ \end{cases} \tag{143}$$

with a large probability of the abstraction reaction. The right column of Fig. 19 shows the electron density of the reaction in the field. (At time $t = 0$ in this figure, pretty much the amount of electrons is already shifted from LiH to H^+.)

Some physical quantities (Mulliken charge, bond lengths, three low-lying adiabatic energies, and their difference) along the present reaction trajectory are depicted for the field-free case in Fig. 20 and for the reaction under laser field in Fig. 21. These two figures clearly illustrate how the laser field dramatically changed the status.

Let us study a little more precisely what kind of difference in the electronic structure is produced by the laser. To do so, we compare the unpaired electron density in Fig. 22. It is very clearly shown that in the field-free case, very large unpaired electron density is observed in H–H during $t = 10$–20 fs. This implies that a stable ground state of hydrogen molecule H–H cannot be produced. On the other hand, due to the presence of the laser, such unpairing is minimized, suggesting that two electrons can share a single molecular orbital to form a hydrogen molecule. This suggests one of the possible guiding principles to control the electronic states. Our next target, which is actually under way, is naturally aimed at a method to create and/or eliminate such bonding status at desired places of a molecular system.

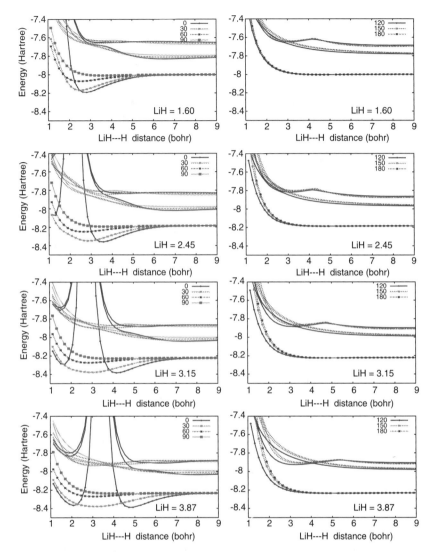

Figure 18. $(LiH_2)^+$ system; adiabatic potential energy curves for the distance between the center of mass of LiH and H with the inserting angle with respect to the LiH and colliding H atom for the LiH length $= 1.60$, 2.45, 3.15, and 3.87 bohrs.

Planar Collisions. Let us collide H^+ with LiH with no impact parameter but different relative angles other than 180° (collinear collision). Place the centroid of LiH on an x-coordinate (Li at left and H at right). The collision angle θ is measured by an angle of the x-coordinate, and the line connecting H^+ and the centroid of

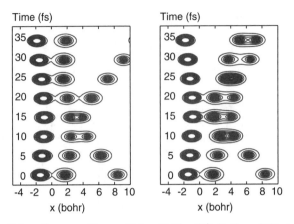

Figure 19.   Effect of laser on collinear collision of LiH + H$^+$. (**Left panel**) Electronically elastic scattering of LiH + H$^+$ in the first excited state without laser. (**Right panel**) Abstraction reaction under the laser field of $E_s = 0.02$ a.u., $\omega = 0.3$ a.u., $t_c = 7.26$ fs, $t_w = 2.42$ fs. The electric field is linearly polarlized along the collision coordinate.

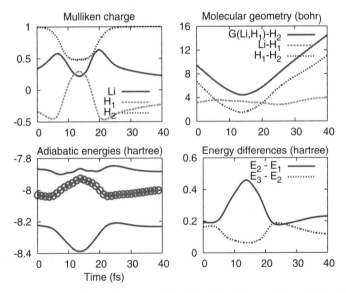

Figure 20.   Variation of some relevant quantities during the collinear collision of LiH + H$^+$ in the first excited state without external field.

LiH (for (LiHH)$^+$ is $\theta = 0°$, while for (HLiH)$^+$ we have $\theta = 180°$). We pick up three configurations: $\theta = 30°$, $60°$, and $90°$. In Fig. 23, we have traced the dynamics of the Mulliken charge (top row), bond lengths (the second raw), bond orders (the third raw), and three low-lying adiabatic energies (the bottom

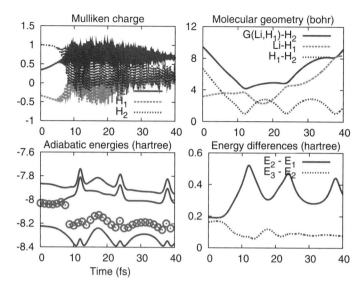

Figure 21. Variation of some relevant quantities during the collinear collision of LiH + H$^+$ in the first excited state under the laser field of $E_S = 0.02$, $\omega = 0.3$, $t_c = 7.26$ fs, $t_w = 2.42$ fs.

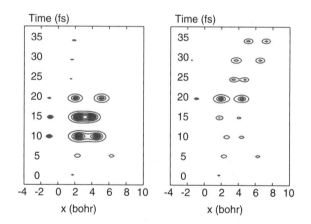

Figure 22. Unpaired electron density to characterize the difference in the collinear collision of LiH + H$^+$ in Fig. 19. (**Left panel**) Electronically elastic scattering of LiH + H$^+$ in the first excited state without laser. (**Right panel**) Abstraction reaction.

raw). The leftmost, the middle, and the right most columns are for the collision angle of $\theta = 30°$, $60°$ and $90°$, respectively.

For these arrangements, we shine the linearly polarized laser in the x-direction as above, but the Gaussian peak time t_c has been selected as 8.96, 12.10, and 13.31

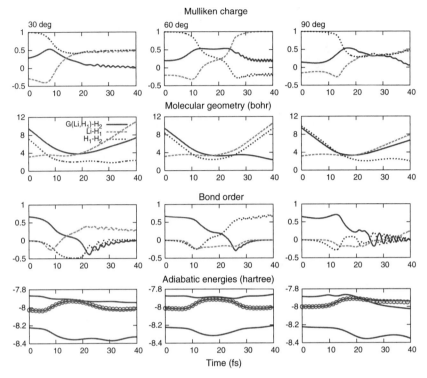

Figure 23. LiH + H$^+$ inserting collision dynamics on the first excited state without external field. The circles in the panel for adiabatic energy denote mean electronic energy.

fs for the angle $\theta = 30°, 60°$ and $90°$, respectively. Figure 24 is the counterpart of Fig. 23 in case of the presence of such laser fields. Three significant differences caused by the laser are (i) vibration excitation of product H–H molecule for all cases, (ii) creation of long-lived (LiH$_2$)$^+$ compound for $\theta = 30°$, and (iii) remarkable change in the reaction channel for $\theta = 60°$. As is seen, the present presentation is never comprehensive to study the entire feature of the reaction. For instance, the polarization vector of the laser field should play an important role. So many parameters should be scanned to make a complete list of reaction feature, and such a study will be reported elsewhere.

H. Control of Nonadiabatic Transition in NaCl

As the last example of the effect of an intense laser on chemical reactions, we study a possible control of nonadiabatic transition of Na + Cl by modulating its avoided crossing.

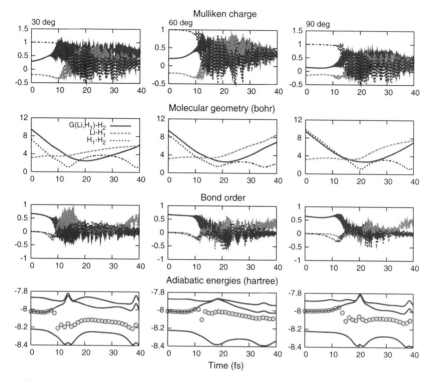

Figure 24. LiH + H$^+$ inserting collision dynamics on the first excited state under the laser field. E_s, ω, and t_w are 0.02, 0.3, and 2.42 fs, respectively. The Gaussian peak times t_c are 8.96, 12.10, and 13.31 fs for the inserting angle $\theta = 30°$, $60°$, and $90°$, respectively.

Figure 25 shows the potential curves that cross each other between the lowest covalent and ionic adiabatic states. The calculation was in the level of RHF/CISD/6-31G, with 26 MO's, 91 CSF's, and 13 core orbitals. We aimed at the qualitatively correct description of only the sigma states, without careful attention to the two lowest Π states. In doing so, we chose the core orbitals up to the second highest occupied molecular orbitals.

The initial distance of two atoms and collision energy was set to 22.67 (bohr) and 1.13415 (hartree) (= 30.86 eV). Two cases of independent asymptotic electronic wavefunctions are studied, namely, the states of Na + Cl and Na$^+$ + Cl$^-$.

The vector potential in this example is a little different from those we used before. It is

$$A(t) = A_s \exp\left(-\left(\frac{t-t_c}{t_w}\right)^2\right)\sin\left(\omega(t-t_c)+\delta\right) \tag{144}$$

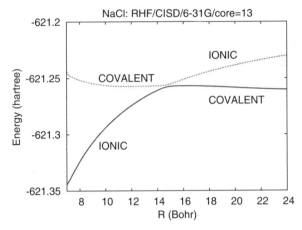

Figure 25. Potential energy curves for the system of Na + Cl around the avoided crossing.

and the associated electric field $E(t)$ is a form of

$$E(t) = \frac{A_s \omega}{c} \exp\left(-\left(\frac{t-t_c}{t_w}\right)^2\right)\left(\frac{2(t-t_c)}{\omega t_\omega^2}\sin\left(\omega(t-t_c)+\delta\right)-\cos(\omega(t-t_c)+\delta)\right)$$

(145)

where $\delta = 0$, $t_c = 20$ (fs), $t_w = 2.42/4.84/7.26$ (fs), $\omega = 0.0017$, and $E_s = 0.02$. The polarization vector is set parallel to the direction of the two atoms. The peak time of laser pulse t_c was chosen as a timing that the atoms come to the minimum energy gap point. We also set the frequency ω so that $\hbar\omega$ should be equal to the minimum energy gap (at this avoided crossing point).

In Fig. 26, we demonstrate a possible laser control of electronic state populations in the present collision dynamics for Na + Cl and Na$^+$ + Cl$^-$. In both cases, the laser width $t_w = 4.84$ fs was found to bring about the largest change of the electronic character. As much as about 30–40% of modulation of the population (the product ratio) has been attained. However, the result is very sensitive to the pulse duration time, as well as to the time scale of nuclear dynamics passing through the avoided crossing region. The guiding principle to control the nonadiabatic transition and the mechanism of it will be discussed in a great detail elsewhere.

VI. CONCLUSION AND PERSPECTIVES

In our age of intense and attosecond pulse laser technology, it is a natural demand to reconsider molecular theory from the viewpoint of an eigenvalue problem in

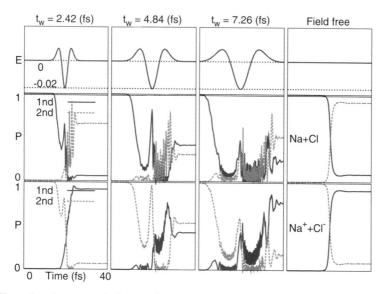

Figure 26. Laser control of electronic state populations for the collisions of Na + Cl (**middle row**) and Na$^+$ + Cl$^-$ (**bottom row**). The electronic populations after crossing the avoided crossing are drawn. The electric fields made by the laser with differnt width are shown in the top row.

the Born–Oppenheimer approximation to that of electron wavepacket dynamics, in which a time coordinate is naturally taken into account. In the latter theoretical scheme, electrons are supposed to synchronize with not only external interactions such as laser fields but also nuclear motions through nonadiabatic interactions. Therefore we have naturally begun this chapter with the standard semiclassical Ehrenfest theory propagated along a nuclear path. In field-free case, nonadiabatic electron wavepackets are propagated along branching nonclassical paths of nuclear motion. These nuclear paths are in turn quantized in terms of nuclear wavepackets running on the individual paths. This establishes a practical way of post-Born–Oppenheimer theory, which we call non-Born–Oppenheimer quantum chemistry, within on-the-fly scheme. The present theory should be useful in many situations where a single isolated potential energy surface is not factored out from other surfaces, in such electronic states that are heavily degenerate.

Having stated the current status of such non-Born–Oppenheimer dynamics in Sections II and III, we considered nonadiabatic electron wavepacket dynamics under laser field in Sections IV and V. An intense laser should induce new kinematic coupling between electrons and nuclei in addition to the native nonadiabatic interactions. Therefore, switching off the laser field corresponds to a situation in which the optically induced nonadiabatic interaction is released and nuclei has left far from a crossing region. However, our study on laser-field nonadiabatic electron wavepacket dynamics is in a little too early a stage to treat it in a full non-Born–Oppenheimer quantum chemistry scheme in contrast to the field-free case. Therefore, we have studied the interplay between laser-field and nonadiabatic interaction in the level of the mean-field approximation (the semi-classical Ehrenfest theory). No ionization manifold in the electronic wavepacket state has been considered.

We have presented some very basic properties of nonadiabatic electron dynamics in laser fields by picking materials from our numerical studies, some of which are of general importance to guide *ab initio* quantum chemistry to the realm of laser-field dynamics. However, due to space limitation, our numerical presentation has been limited to the level of preliminary exploration. More systematic and thorough studies and analyses will be reported elsewhere.

Acknowledgments

This work has been supported in part by the Grant-in-Aid from the Ministry of Education, Culture, Sports, Science and Technology of Japan.

References

1. M. Born and R. Oppenheimer, *Ann. Phys.* **84**, 457 (1927).
2. S. Takahashi and K. Takatsuka, *J. Chem. Phys.* **124**, 144101 (2006).
3. L. D. Landau, *Phys. Zts. Sov.* **2**, 46 (1932).
4. C. Zener, *Proc. R. Soc. London* **137**, 696 (1932).
5. E. C. G. Stueckelberg, *Helv. Phys. Acta* **5**, 369 (1932).
6. P. Pechukas, *Phys. Rev.* **181**, (1969). 174
7. W. H. Miller and T. F. George, *J. Chem. Phys.* **56**, 5637 (1972).
8. W. H. Miller and C. W. McCurdy, *J. Chem. Phys.* **69**, 5163 (1978).
9. C. A. Mead and D. G. Truhlar, *J. Chem. Phys.* **77**, 6090 (1982).
10. C. Zhu and H. Nakamura, *J. Chem. Phys.* **102**, 7448 (1995).
11. C. Zhu and H. Nakamura, *J. Chem. Phys.* **106**, 2599 (1997).
12. X. Sun and W. H. Miller, *J. Chem. Phys.* **106**, 6346 (1997).
13. G. Stock and M. Thoss, *Phys. Rev. Lett.* **78**, 578 (1997).
14. K. Takatsuka and H. Nakamura, *J. Chem. Phys.* **85**, 5779 (1986).
15. J. C. Tully, *J. Chem. Phys.* **93**, 1061 (1990).
16. C. Zhu, S. Nangia, A. W. Jasper, D. G. Truhlar, *J. Chem. Phys.* **121**, 7658 (2004).
17. C. Zhu, A. W. Jasper, D. G. Truhlar, *J. Chem. Phys.* **120**, 5543 (2004).
18. M. D. Hack, A. M. Wensmann, D. G. Truhlar, M. Ben-Nun, and T. J. Martìnez, *J. Chem. Phys.* **115**, 1172 (2001).
19. A. W. Jasper, M. D. Hack, A. Chakradorty, D. G. Truhlar, and P. Piecuch, *J. Chem. Phys.* **115**, 7945 (2001).
20. R. Baer, D. J. Kouri, M. Baer, and D. K. Hoffman, *J. Chem. Phys.* **119**, 6998 (2003).
21. M. Ben-Nun and T. J. Martìnez, *J. Chem. Phys.* **108**, 7244 (1998).

22. E. R. Bittner and P. J. Rossky, *J. Chem. Phys.* **103**, 8130 (1995); *ibid* **107**, 8611 (1997).
23. K. F. Wong and P. Rossky, *J. Chem. Phys.* **116**, 8418 (2002); *ibid* **116**, 8429 (2002).
24. S. Hammes-Schiffer and J. C. Tully, *J. Chem. Phys.* **101**, 4657 (1994).
25. M. F. Herman and M. P. Moody, *J. Chem. Phys.* **122**, 094104 (2005).
26. P. Oloyede, G. Mil'nikov, and H. Nakamura, *J. Chem. Phys.* **124**, 144110 (2006).
27. V. Krishna, *J. Chem. Phys.* **127**, 134107 (2007).
28. N. Ananth, C. Venkataraman, and W. H. Miller, *J. Chem. Phys.* **127**, 084114 (2007).
29. J. R. Schmidt and J. C. Tully, *J. Chem. Phys.* **127**, 094103 (2007).
30. C. M. Isborn, X. Li, and J. C. Tully, *J. Chem. Phys.* **127**, 134307 (2007).
31. H. D. Meyer and W. H. Miller, *J. Chem. Phys.* **70**, 3214 (1979).
32. M. S. Child, *Molecular Collision Theory*, Academic Press, New York, 1974; *Semiclassical Mechanics with Molecular Approximations*, Clarendon Press, Oxford, 1991.
33. H. Nakamura, *Nonadiabatic Transition*, World Scientific, Singapore, 2002.
34. A. Jasper, B. K. Kendrick, C. A. Mead, and D. G. Truhlar, *Modern Trends in Chemical Reaction Dynamics, Part I*, X. Yang and K. Liu, editors, World Scientific, 2004, Chapter 8.
35. M. Baer, *Beyond Born–Oppenheimer*, Wiley, Hoboken, NJ, 2006.
36. H. Ushiyama and K. Takatsuka, *Angew. Chem. Intl. Ed.* **46**, 587 (2007).
37. M. Amano and K. Takatsuka, *J. Chem. Phys.* **122**, 084113 (2005).
38. F. Krausz and M. Ivanov, *Rev. Mod. Phys.* **81**, 163 (2009).
39. M. Hentschel, R. Kienberger, C. Spielmann, G. A. Reider, N. Milosevi, T. Brabec, C. Corkum, U. Heinzmann, M. Drescher, and F. Krausz, *Nature* **414**, 509 (2001).
40. H. Niikura, F. Légaré, R. Hasban, A. D. Bandrauk, M. Y. Ivanov, D. M. Villeneuve, and P. B. Corkum, *Nature* **417**, 917 (2002).
41. Y. Mairesse, A. de. Bohan, L. J. Frasinski, H. Merdji, L. C. Dinu, P. Monchicourt, P. Breger, M. Kovacev, T. Auguste, and B. Carrè et al., *Phys. Rev. Lett.* **93**, 163901 (2004).
42. Y. Nabekawa, T. Shimizu, T. Okino, K. Furusawa, K. Hasegawa, K. Yamanouchi, and K. Midorikawa, *Phys. Rev. Lett.* **96**, 083901 (2006).
43. J. J. Carrera, X. M. Tong, and S. I. Chu, *Phys. Rev. A* **74**, 023404 (2006).
44. Z. Hatsagortsyan and C. H. Keitel, *Phys. Rep.* **427**, 41 (2006).
45. A. Baltuška, T. Udem, M. Uiberacker, M. Hentschel, E. Goulielmakis, C. Gohle, R. Holzwarth, V. S. Yakovlev, A. Scrinzi, A. Hansch, et al., *Nature* **421**, 611 (2003).
46. E. Deumens, A. D. H. Taylor, and Y. Öhrn, *J. Chem. Phys.* **96**, 6820 (1992).
47. E. Deumens A. D. R. Longo, and Y. Öhrn, *Rev. Mod. Phys.* **66**, 917 (1994).
48. J. Broeckhove, M. D. Coutinho-Neto, E. Deumens, and Y. Öhrn, *Phys. Rev. A* **56**, 4996 (1997).
49. M. Elshakre and F. Kong, *Trends Chem. Phys.* **10**, 179 (2002).
50. I. P. Kràl and M. Shapiro, *Rev. Mod. Phys.* **79**, 53 (2007).
51. I. Kawata, H. Kono, and Y. Fujimura, *J. Chem. Phys.* **110**, 11152 (1999).
52. M. Uhlmann, T. Kunert, F. Grossmann, and R. Schmidt, *Phys. Rev. A* **67**, 013413 (2003).
53. K. Yagi and K. Takatsuka, *J. Chem. Phys.* **123**, 224103 (2005).
54. P. Krause, T. Klamroth, and P. Saalfrank, *J. Chem. Phys.* **123**, 074105 (2005).
55. T. Klamroth, *J. Chem. Phys.* **124**, 144310 (2006).
56. H. B. Schlegel, S. M. Smith, and X. Li, *J. Chem. Phys.* **126**, 244110 (2007).
57. M. Baer, *J. Phys. Chem. A* **110**, 6571 (2006).

58. B. Sarkar, S. Adhikari, and M. Baer, *J. Chem. Phys.* **127**, 014301 (2007).

59. S. Ramakrishna and T. Seideman, *Phys. Rev. Lett.* **99**, 103001 (2007).

60. S. S. Viftrup, V. Kumarappan, S. Trippel, H. Stapelfeldt, E. Hamilton, and T. Seideman, *Phys. Rev. Lett.* **99**, 143602 (2007).

61. J. H. Posthumus, *Rep. Prog. Phys.* **67**, 623 (2004).

62. I. Last and M. Baer, *J. Chem. Phys.* **82**, 4954 (1984).

63. A. Giusti-Suzor, F. H. Mies, L. F. DiMauro, E. Charron, and B. Yang, *J. Phys. B* **28**, 309 (1995).

64. N. A. Nguyen and A. D. Bandrauk, *Phys. Rev. A* **73**, 032708 (2006).

65. K. Harumiya, I. Kawata, H. Kono, and Y. Fujimura, *J. Chem. Phys.* **113**, 8953 (2000).

66. K. Takatsuka, *J. Chem. Phys.* **124**, 064111 (2006).

67. K. Takatsuka, *J. Phys. Chem. A* **111**, 10196 (2007).

68. A. Jasper, B. K. Kendrick, C. A. Mead, and D. G. Truhlar, *Modern Trends in Chemical Reaction Dynamics*, Part I, X. Yang and K. Liu, editors, World Scientific, Singapore, 2004.

69. W. Domcke, D. R. Yarkony, and H. Köppel, *Conical Intersections: Electronic Structure, Dynamics & Spectroscopy*, World Scientific, Singapore, 2004.

70. J. C. Tully, *Faraday Discuss.* **110**, 407 (1998).

71. R. Kapral, *Annu. Rev. Phys. Chem.* **57**, 129 (2006).

72. T. Yonehara and K. Takatsuka, *J. Chem. Phys.* **129**, 134109 (2008).

73. S. Takahashi, M. Fujii, and K. Takatsuka, submitted for publication (2009).

74. K. Takatsuka and A. Inoue-Ushiyama, *Phys. Rev. Lett.* **78**, 1404 (1997); A. Inoue-Ushiyama and K. Takatsuka *Phys. Rev. A* **59**, 3256 (1999).

75. S. Takahashi and K. Takatsuka, *J. Chem. Phys.* **127**, 084112 (2007).

76. S. Schulman, *Techniques and Applications of Path Integration*, Wiley, New York, 1981.

77. T. Yonehara, S. Takahashi, and K. Takatsuka, *J. Chem. Phys.* **130**, 214113 (2009).

78. T. Yonehara and K. Takatsuka, *J. Chem. Phys.* **128**, 154104 (2008).

79. G. C. Schatz and M. A. Ratner, *Quantum Mechanics in Chemistry*, Prentice-Hall, Englewood Cliffs, NJ, 1993.

80. A. D. Bandrauk, E.-W.S. Sedik, and C. F. Matta, *J. Chem. Phys.* **121**, 7764 (2004).

81. M. P. Allen and D. J. Tildesley, *Computer Simulation of Liquids*, Oxford University Press, Oxford, 1987.

82. H. Tal-Ezer and R. Kosloff, *J. Chem. Phys.* **81**, 3967 (1984).

83. L. I. Schiff, *Quantum Mechanics*, McGraw-Hill, New York, 1968.

84. E. Steiner and P. W. Fowler, *J. Phys. Chem. A* **105**, 9553 (2001).

85. M. Tsukada, K. Tagami, K. Hirose, and N. Kobayashi, *J. Phys. Soc. Jpn* **74**, 1079 (2005).

86. I. Barth, J. Manz, Y. Shigeta, and K. Yagi, *J. Am. Chem. Soc.* **128**, 7043 (2006).

87. I. Barth, H.-C. Hege, H. Ikeda, A. Kenfack, M. Koppitz, J. Manz, F. Marquardt, and G. K. Paramonov, *Chem. Phys. Lett.* in press.

88. M. Okuyama and K. Takatsuka, *Chem. Phys. Lett.* **476**, 109 (2009).

89. K. Takatsuka, K. Yamaguchi, and T. Fueno, *Theor. Chim. Acta* **48**, 175–183 (1978).

LIQUID BILAYER AND ITS SIMULATION

J. STECKI

Department III, Institute of Physical Chemistry
Polish Academy of Sciences, Warsaw, Poland

CONTENTS

I. INTRODUCTION

Bilayers are just one of the many enormously varied structures formed by amphiphilic molecules. Because of the importance of physics and chemistry of membranes, lipid bilayers are objects of great and extensive interest. These topics are vast with many ramifications.

A molecule is called amphiphilic if it contains two disparate parts—a hydrophilic part and a hydrophobic part—joined by a permanent chemical bond.

Advances in Chemical Physics, Volume 144, edited by Stuart A. Rice
Copyright © 2010 John Wiley & Sons, Inc.

A hydrophilic group [1–6] is polar with strong affinity to water and is called the "head." A hydrophobic group is typically made of aliphatic hydrocarbon chains. These two groups of opposing affinities are forced to travel together through space and time, so the amazingly intricate and beautiful structures are formed [4, 9] in order to accommodate this. Many of these are surfactants, and very often "surfactant" and "amphiphilic" are used interchangeably. A small subgroup of such compounds is comprised of lipids [1–4]. These all-important compounds form or enter membranes and in many ways are present in living matter [2, 4]. A small subgroup of structures—spatial structures—formed by the lipids consists of bilayers and vesicles. Vesicles are of various shapes (such as spherical), and they enclose a finite volume (filled with some fluid, such as water or aqueous solution). This distinguishes vesicles from bilayers: A bilayer does not enclose a finite volume "inside" because it does not have an "inside." A bilayer is nearly planar; so instead of "inside" and "outside" of a vesicle, it has an "upper" side and a "lower" side, with both sides ideally in contact with the same solvent. If so, it has no spontaneous curvature expected in a vesicle. In a solvent, usually water, the amphiphilic molecules form micelles [4]. These objects of various shapes (cylindrical, spherical, etc.) are well known to physical chemistry and chemical physics [6], although a quantitative description proved elusive and all we have are superficial notions—which is not surprising given the ill-defined object and such objects the micelles are. A bilayer in this context may be thought of as a giant micelle that chooses to be planar instead of another shape. The more ambitious simulations demonstrate the process of such self-assembly.

Here we limit the discussion to planar bilayers, leaving aside vesicles and micelles. We also limit the scope, focusing on simulation of the simplest model amphiphiles in *a liquid* bilayer. General Bibliography [1–9] includes introductions to this exotic field [1–4]; theory of all inter faces [8] and their thermodynamics [6–8], plus the background of the intermolecular forces [4–6].

II. DEVELOPMENT OF SIMULATIONS

The very first succesful simulation of a liquid bilayer was published in 1988 [10] and used a model of amphiphilic dimers: One end represented the hydrophilic head while the other little sphere represented the hydrophobic tail, connected by a permanent bond. This was very soon extended and generalized [11]. Since then a great profusion of work appeared developing along distinct lines, roughly along two avenues. In one the aim was to realistically represent (i.e., model as accurately as possible) the large organic molecules (i.e., lipids) with all aspects of inter- and intramolecular forces including the long-range electrostatic inter-actions. The latter originate with the polar solvent (i.e., water) and with the polar heads of the amphiphile. Eventually, ready packages of dedicated software became available. Even with the supercomputers available in the 1990s, simulation of more than 64–1024 lipid molecules was at the limit of the possible. Since then an overwhelming variety of work appeared dealing with methodical improvements and dedicated to quasi-realistic modeling of large molecules and processes entirely directed to biological processes and systems and to prospective medical applications as well. A few representative references (out of hundreds if not thousands) might give an idea of the obstacles and computational requirements [12–53]. Alternatively, the other approach developed in which the molecular model was simplified to the extreme [10, 11]. In part, the aim was also modified and often was not so much as "to simulate" as realistically as possible but rather *to use* the simulation as a tool for understanding the physical laws of these special systems—that is, to find out "what makes them tick."

Such a greatly simplified model, which nevertheless became widely accepted, uses any simple potential function for the description of intermolecular forces, which models steric repulsion and dispersion-force attraction. Most often it will be the Lennard-Jones 6–12 potential. It is taken to act between the monomeric segments of which the chain-like tails of the amphiphile are built, as well as to act between heads, between the pairs head–tail segment, between the pairs of solvent molecules, also spherical, and between the solvent molecules and every element of which the bilayer is constructed. The arguments for such simplification are easy to see; one can simulate much bigger systems, even with thousands of pseudo-lipids in a solvent. Also, in a simplified system, one has control of the much less numerous parameters (just a few), and one can gain an understanding of the interplay of various effects.

With increased computer power and with simplifications of the model used, much bigger systems could be simulated over much longer times. Then mesoscopic effects began to show up, complicating the picture. The most common of these were undulations, which could not develop in a small system over a short time. These physically important "capillary waves" are discussed in Section 4. But it has to be said that for many studies the undulations are just an unwanted complication that one wishes were not there at all; therefore for some specific questions (like phase behavior of chains) it is advantageous to simulate small systems on purpose, not extending simulation time too much either [23–26]. Then undulations do not have enough time to develop. Two papers belonging to the group of "quasi-realistic" models are still classical references to structure factors of undulations [16–18]. Peristaltic motions were also described quantitatively [16–18].

We can today see very well the substantial and positive influence the simulations have had on general understanding. The early theoretical work contained much of what became improved, not to mention the erroneous.

Simulations helped greatly to establish the true image and led to quite several rectifications.

We describe now in turn: the model(s), the method(s), and the quantities.

But we begin with the quantities that one can (and does) determine—the most interesting topic of all.

Each and every simulation routinely produces averages of such quantities as kinetic temperature T, the internal energy U, and pressure p; in fact, anything that can be expressed as a function of positions and momenta of particles making up the system. Thus entropy or thermodynamic temperature cannot be had. Many special methods have been devised in order to break out of this limitation. Also different statistical ensembles have been used.

Most obviously a simulation of a planar bilayer, which is a two-dimensional object embedded in three dimensions, determines the "surface tension." Borrowing from surface physics and chemistry (see, e.g., Refs. 6, 7, and 8), one asks the obvious question about the surface tension. Next as a topic comes the corresponding area. Then, the surfaces and interfaces are known to execute strong shape fluctuations, so one asks for correlations and any other means of describing quantitatively the shape fluctuations of the bilayer as a whole and of each monolayer separately.

Since the system in the simulation box—the planar bilayer in the solvent or in the vacuum created by the simulator—is anisotropic, the pressure becomes a tensor and one asks for the nonzero elements, namely, the normal $p_N = p_{zz}$ and the tangential $p_T = p_{xx} = p_{yy}$. Even more, one can take the definition of the local pressure tensor $p_{\alpha\beta}(x, y, z)$ and ask for the "profile" $p_N(z)$ and $p_T(z)$. Luckily, for this symmetry in a rectangular simulation box $L_x \times L_y \times L_z$, the profiles are well-defined.

A novel global property is the radius of gyration. Initially defined for polymers, it has been discussed for membranes as well.

Next we ask for more varied information; as the bilayer is made of anisotropic chain molecules, one can determine the average orientation of the first bond linking the "head" and the chain-like "tail", for example. Even the distribution of orientations can be had. The correlation between two such bonds can also be determined and can be linked to the correlations of normals to the surface representing the bilayer.

Once the focus is on the chains making up the "tail" in the "head-and-tail" model of the amphiphilic molecule, a plethora of very detailed and specific questions appears which were and are central to physics and chemistry of lipids. Whether the simulator can contribute anything depends very much on how realistic his model is. In the simplest case of a dimer the answer is simple: almost nothing. For hydrocarbon-like tails modeled as a freely jointed chain, the possibilities are better: the end-to-end distance, the individual radius of gyration, and the "interdigitation", meaning the interpenetration of a chain into the "other" monolayer. An improved model introduces some rigidity of the hydrocarbon chain. All "realistic" models introduce angle-dependent potential functions for the configurations of the successive bonds; then the trans and cis configurations become distinct. In this review the focus is on the simplified models and on general laws and concepts.

The determination of the "true" area of the bilayer is a special topic we discuss in Section IX.

Besides the lateral tension as a thermal average, the full probability distribution for it can be determined from its histogram.

Finally, the shape fluctuations are described in terms of the structure factor $S(q)$, and its interpretation occupy a section of its own (Section X).

Turning now to the brief description of methods; References 19 and 20 describe the methods and technicalities of the computer simulation. Bilayers belong to systems obeying classical mechanics. Reference 20 is a newest advanced text. A discussion of time discretization is interesting [21, 22].

Essentially there is the Monte-Carlo method and the Molecular Dynamics (MD) method. Both need the positions of particles, and MD also deals with velocities as it solves the Newton equations of motion. For that reason, MD does not know energies, only forces. Truly a closed system does not raise questions, but constant temperature is already an ambiguous affair because the thermodynamic temperature is not known (only the kinetic temperature is known) and how to keep it constant is a relatively recent invention under the name of Nose–Hoover thermostat [19, 20]. Therefore we rely on the proofs of equivalence for the kinetic temperature for a classical system at equilibrium.

The relatively new method of Dissipative Particle Dynamics (DPD) offers serious advantages in shortening the computation time and producing results in relatively small systems; an excellent detailed description is found in Ref. 23, followed by extensive results [23–26]. But DPD has also been not only used [27, 28] but criticized [28]. The problem with DPD is the arbitrariness of the parameters specifying the effective interactions; the "particle" is now something like a unit volume comprising a few objects that would be beads in the atomic picture. Therefore a non-negligible amount of effort went into gaining experience, and essentially the parameters are adjusted so as to produce the expected or experimentally known results—like sequences of phases of bilayers in a range of temperatures [23]. But now this knowledge has been accumulated, and presently the interest of simulators has shifted to more complicated systems (above all a membrane with a protein embedded) or to events, such as opening a pore (see Section III) or membrane fusion, and so on.

The main reservation one might have against this kind of work is the extraordinary arbitrariness of Hamiltonians, potentials, and potential energy parameters, which shock a physical chemist's elementary knowledge of intermolecular forces.

But much of this work is in the hands of theoretical physicists, for whom nothing is sacred and who have nucleon–nucleon delta-function attraction for breakfast.

Total arbitrariness of the potentials changed at will to fit the experiment—certainly this kind of thinking comes from nuclear physics and is alien to chemical physics or physical chemistry. The reason is very simple: Intermolecular forces are *known* and even known in detail.

Atomistic simulations (as they are called now) began with early work quoted above [10, 11] and was later taken up by the Max Planck group [29–35] and others [36–46]. These will be described in Sections III and IV. But soon the DPD method took over [29–33]. The rare atomistic simulations [34, 37–40, 45] examined series of areas at constant temperature and density (see Section III). Also the bilayer isotherm was partly determined by the DPD method [29–33] and by atomistic simulation with special potentials [46].

A serious amount of effort went into designing solvent-free models [47–54], the motivation being the inordinate amount of computer time spent on the solvent particles. The latter provide only an external field of sorts and their presence is secondary. One apparently can obtain a stable bilayer in vacuum, although the interactions have to be adjusted rather arbitrarily if the bilayer is to stay and not to disperse. A particular variant introduces a "phantom solvent" instead [53, 54]. There one also finds criticism of the weaknesses of the DPD method.

This brings us to the topic of details of the molecular model. Whether in atomistic or in DPD simulation, the amphiphile is modeled as a chain of beads. One bead is to represent the (normally polar) head, and the rest of the chain is to represent the (hydrophobic hydrocarbon) tail. The solvent (normally water) is made of spherical particles. The possibilities are overwhelming: one tail or two (as suggested by real lipids), sizes of beads all the same (preferred), interactions between beads (many), and all pairs made of s (solvent), a* (head), or b* (a bead from the tail plus intramolecular properties of the permanent bonds). On top of that, the tail may be invested with some rigidity—for example, in the form of the three-body potential between three successive b*'s [34]. But that is not enough because water is polar; that is, its interactions depend on orientation, *and* it is the source of the *hydrophobic effect* which is responsible for keeping the tails away from the solvent (i.e., inside the bilayer). This effect has been included by adding an extra "deus ex machina" repulsive potential between s and b* (r^{-9}) or similar. Keeping things simple—that is, all sizes equal and all pair interactions equal, (with the addition of rigidity + hydrophobicity)—led to a success [34, 35]. Not only was a stable bilayer with molecules of length $\ell = 4$ simulated over a range of areas (the bilayer isotherm; see Section III), but also the bilayer did self-assemble from isolated molecules in the solvent.

As a sequel to that work, the first ever $1/q^4$ divergence was found in a simulation [35].

The arbitrary repulsive potential that replaces the action of hydrophobic forces can be dispensed with in a rather special model, which, however, worked. The same amphiphiles that form the bilayer will also form micelles of many shapes. The inspiration comes from the existence of *reverse micelles* in which the heads hide in the center and the hydrophobic tails go to the outside core—in a *nonpolar solvent*. Ordinarily, moving along the z-axis we have solvent–heads–tails–tails(again)–heads–solvent—or, in terms of the cohesive energy [5] density, high–high–low–low–high–high. Now in reverse bilayers in a nonpolar solvent the coehsive energy density is low(solvent)–low(hydrocarbon tails)–high(heads)–high(heads)–low(tails)–low(solvent). Having a nonpolar solvent, we do not have any qualms representing it by spherical particles and we do not have to introduce an extra potential to mimic the hydrophobic effect. This was done [36, 37, 39] (see also Ref. 40) for the shortest chain $\ell = 1$ (i.e., for a dimer). With all sizes equal, the dimer made of two spheres (close but not fused) keeping all interaction energies equal, except for the a^*a^* head–head attraction seriously bigger, one has successful formation of a *reverse bilayer* [36, 37, 39, 40]. For this system, many properties and correlations were reliably determined—some will be quoted in subsequent sections. However, it has to be said that the extra repulsion coming from the solvent was used in some series because then the equilibrium was reached faster and the overall behavior of the simulated bilayer was more steady and stable.

The objections one might raise against this model have more to do with thermodynamics (see, e.g., Refs. 6 and 8): now the formation of the bilayer is predominantly energetic. In normal situations with water as solvent, entropic effects are much more significant. But such physical chemistry has not been tackled yet.

Whereas for chain length $\ell = 1$ the "non-reverse" bilayer would not be stable if all energy parameters were kept equal (with or without the hydrophobic extra), for chain length $\ell = 4$ and $\ell = 8$ our simulations were without any extra attraction [38, 40]. Apparently the steric effects were enough to keep the bilayer together and the amphiphiles inside, tail-to-tail with heads facing the solvent, once the bilayer was constructed.

Besides molecular models used by the simulator, there is also a subvariety of "theoretician's models." The most important one is the model of continuous surface, or (rarely) two surfaces as mathematical objects to which differential geometry can be applied. If, moreover, the surface in question does not depart too much from the flatness of the plane x–y, the surface can be represented by a single-valued function $z(x, y)$. When discussing or determining the undulations, these notions pervade all calculation and interpretation (Section IV). More general cases of mathematical surfaces are also covered by the celebrated Canham–Helfrich hamiltonian [55] (see Section IV).

III. LATERAL TENSION

As mentioned in Section II, the first and prominent result of a simulation is the lateral tension, Γ. It is coupled to the area provided by the (rectangular) simulation box. We quote in abbreviated form the best derivation [8, p. 89] of the working equations for it.

The simulation box of volume $V = L_x L_y L_z$ contains the bilayer and the solvent. The "black-box" derivation of the exact expression for the lateral tension assumes only that the volume V is filled with some particles, interacting with some conservative potential, and that classical mechanics applies. Thus the free energy $F(T, V, N)$ really is $F(T, L_x, L_y, L_z; N)$ and we borrow from thermodynamics the hint to consider F as $F(T, V, A, \ldots)$, where A is the area. Therefore we wish to calculate the partial derivative of F with respect to the area A at constant volume V. Let us take $A \equiv L_x L_y$, and let us expand the box to $L_x(1 + \xi), L_y(1 + \eta)$, $L_z(1 + \zeta)$; to first order $\Delta V/V = \xi + \eta + \zeta$, which we wish to annul. Therefore we must take $\zeta = -\xi - \eta$; the box is compressed along the z-axis and expanded along the other two; $\Delta A/A = \xi + \eta$. The free energy, also to first order, is obtained as a sum of three terms calculated by the well-known recipe of rescaling the coordinates in the configurational part of the partition function. The rescaling produces forces as derivatives of total energy and we end up with a simple expression

$$\Gamma A = (2P_{zz} - P_{xx} - P_{yy})/2 \tag{1}$$

where P's are the diagonal elements of the pressure tensor. All is well-defined for a parallelepiped. If the system happens to be homogeneous, we will get null. Conversely, any inhomogeneity in the z-direction may produce a nonvanishing result, but whether it will have the desired meaning, or any meaning at all, is not guaranteed. It is truly a black-box formula.

The P's are averages of moments of forces

$$P_{\alpha\alpha} = (-)\left\langle \sum_j r_{j,\alpha} \partial U/\partial r_{j,\alpha} \right\rangle, \qquad \alpha = x, y, z \tag{2}$$

where U is the potential energy of the black-box particles and the index j refers to particles. Kinetic contributions cancel out and were not included in the quotation of these well-known expressions. As always, statistical mechanics produces pV and ΓA.

The derivation was first given a long time ago and is explained and reproduced in many places. We have quoted the argument because there were some doubts variously expressed about the validity of the Kirkwood–Buff derivation and/or its applicability to membranes or bilayers. Doubts must have originated with those theoretical physicists who have publicly stated that "Kirkwood set back statistical mechanics for 50 years," no doubt wishing to solve everything with a Landau theory polynomial. No, the doubts were misplaced, the black-box may contain a bilayer as well. In the derivation, either free energy (in the Gibbs canonical ensemble) or other thermodynamic potentials may be used [8, 56].

Simulators have generally included the lateral tension in the computed averages, calling it sometimes by other names, but I was unable to trace the origin of the adjective "lateral" with certainty. I attribute it provisionally to the Max Planck group [29–35]. It is a good term because of the firm distinction from other tensions. Γ is of course coupled to $A = A_\perp = L_x L_y = L^2$ in our notation, the edge of the box, or the "projected" area of the fluctuating bilayer, as evidenced by the derivation sketched above. More about "projected" area in Section V.

Most simulations were limited to the "tensionless state," a state in which the bilayer, membrane, or vesicle is "at rest" without external forces acting on it. It is a convenient reference state of room temperature and pressure $P = 1$ atm, suggested to simulator by the experiment in the laboratory. The tensionless state in simulations is usually identified with zero lateral tension, in fact in all published work with the exception of Refs. 32 and 33. However, there are other tensions that can be defined that we refer to in Section IV, devoted to the structure factor, and in Section V, devoted to the small-gradient theory. Then other tensionless states result.

Beginning with early attempts, followed by a larger interval of the bilayer isotherm [21], more work appeared in which a series of states with a series of areas were simulated, of course at the same temperature but also at the same number of amphiphiles N and the same global density $\mathcal{N}/V$ [29–35, 37–39, 42–45]. At these conditions, one can still change the shape of the box and therefore have a series of areas $A_\perp$ and $a = 2A_\perp/N$. Simulations that covered the entire area of stable existence of the bilayer are relatively rare [37, 38, 45, 46].

A resulting function of Γ versus $a = 2A_\perp/N_s$, the area per head, has many distinct and interesting features. It may be termed *the bilayer isotherm*. Such typical isotherms are shown in Figures 1–6.

Figure 1 shows three isotherms for very small systems, which show that particular feature of discontinuous change in slope. This is not seen in systems that are just a little larger.

Generally in the region of high a the lateral tension increases with a linearly until the bilayer breaks under tension. But what makes it entirely different from any surface/interfacial tension, is that it depends on the area at all!

Lowering the area, we lower Γ as well, cross the axis (at the lateral-tensionless state $a = a_0$), and continue smoothly into the negative Γ's. So not only Γ depends

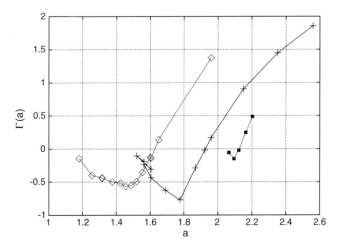

Figure 1. Three bilayer isotherms, two fuller and one a small interval, plotted as lateral tension Γ against area per head a for truly small systems; note continuation of the isotherm deep into the region of negative Γ and the sudden change of the slope (i.e., of the derivative $d\Gamma/da$). $N = 450.\ell = 4, 8, 4$. Black boxes: data from Ref. 46.

on the area, but also it can get negative. Further details depend on the size of the system.

We call the region of high a, positive Γ the "EX" or "extended" bilayer, because the bilayer there is almost flat, gently undulating, and extended compared to the tensionless state, at least if $\Gamma > 0$. The second derivative is negative.

The tensionless state does not appear on the isotherm as anything special—just the curve crosses the x-axis of a. Then on further compression, a small system enters the *floppy* region with a large jump in the slope $d\Gamma/da$ which changes sign from positive to negative. The three typical bilayer isotherms for truly small systems are shown in Fig. 1. The portion to the right of the figure, that of larger a's, has been reported many times [29–34, 36–40], even if only a part of the bilayer isotherm was determined [29–34]. In most, if not all, the second derivative of Γ versus a was slightly negative. The exception was reverse dimers [37, 39]. The abrupt change of slope marks the transition from Extended to Floppy. "Floppy," "fuzzy," "foamy," and "buckling" bilayer, this is how it was described. Visualization of the positions of particles shows (in, e.g., Refs. 37, 45, and 46) indeed a very different picture from that seen in the region "EX" of extended bilayer. The bilayer is still there, it has not disintegrated, but it a very chaotic structure extruding (buckling) and, in the end (i.e., at smallest a's), penetrated partly by the solvent. Compressed laterally still further, it disintegrates in various ways—for example, by changing into a "quad-layer" still of indeterminate shape.

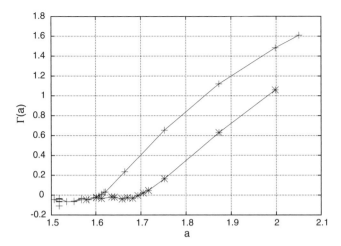

Figure 2. Effect of temperature: two bilayer isotherms for fairly large systems (note plateau for low a); always higher T shifts the entire isotherm to the right—that is, to larger a. Therefore at const a, with T rising EX → FL. $N \simeq 4000$.

Such last data points were not reported in the published work. The descriptions in words have been given several times [33, 37–40, 45, 46] by the simulators. That is so because we do not have an order parameter whose value would tell us if we are in the FL region—as is customary in the field of phase transitions. Therefore we have to resort to descriptions.

Figure 2 shows the effect of the temperature: two bilayer isotherms for the same bilayer but at two temperatures. A shift to the right (to higher areas) with increased temperature means that extended bilayer from the region EX when heated will transform itself into a floppy form. But this happens in a limited interval of areas, essentially between two tensionless states, in this example from $a \simeq 1.61$ to $a \simeq 1.7$.

After the appearance of the "floppy" state, a next surprise comes when one examines the size dependence. Since the earliest days of computer simulation, the main challenge and also the main line of criticism was the smallness of the simulated system. To increase the number of particles $\mathcal{N}$ and thereby the volume and/or area became the challenge, sometimes even the main challenge. Besides, several values of $\mathcal{N}$ are needed to do the necessary extrapolation to macroscopic size and thermodynamic limit. It is therefore the most obvious task, but for simulated bilayers the first paper that did compare several sizes appeared only in 2004 [38]. What happens is illustrated by the plots of Figs. 3, 4, and 5.

First, the EX region of high a is not changed at all; the lateral tension $\Gamma(a)$ does not depend on N there. There is a common line for all sizes.

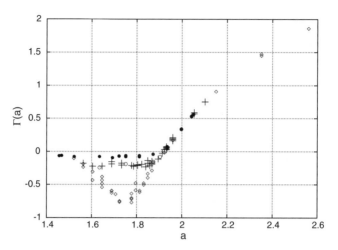

Figure 3. Effect of size on a complete bilayer isotherm from hole at $a \simeq 2.9$ to disintegration at $a \simeq 1.35$; Diamonds, $N = 450$; plus signs, $N = 1800$; circles, $N = 4000$ (open circles—metastable); $\ell = 8$, $T = 1$.

Then in the region of negative Γ, with the exception of the very small systems (such as shown in Fig. 1), the isotherm becomes a flat plateau of negative Γ very weakly depending on a, if indeed at all. Only the very small systems do not show the plateau, and the reverse dimers [39] did show plateau's for all sizes. The plateau is flatter and $|\Gamma|$ smaller if the system is bigger. Visualization does not bring

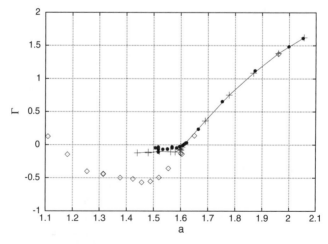

Figure 4. See caption to Fig. 3. A shorter chain $\ell = 4$ easier to equilibrate [6, 42–46]; a complete isotherm [33] supplemented with new data [61].

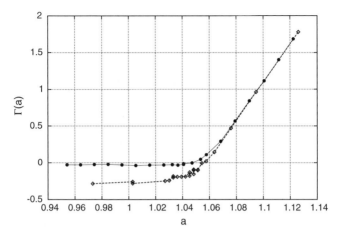

Figure 5. Compare to Figs. 3 and 4. Reverse dimers. Diamonds, $N = 2880$; circles, $N = 18,720$, $T = 1.9$.

anything new: The bilayers look very much the same for all sizes—"floppy," "foamy," or "fuzzy" as described above.

Disposing of the scatter of the experimental points and omitting the very smallest systems, we can schematically summarize the situation by Fig. 6: linear dependence of $\Gamma(a)$ and plateaus of size-dependent Γ-floppy.

The plateau value of $|\Gamma|$ in the floppy region was found [39] to scale as $1/N$ at fixed a; this means comparing different systems of different sizes, N being the

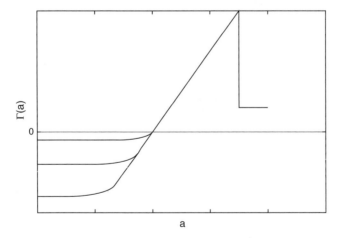

Figure 6. Scheme suggested by the data on full bilayer isotherms: Three sizes produce three different plateaus in the FL floppy region and produce a common $\Gamma(a)$ in the extended EX region. The isotherm terminates with a sudden drop on appearance of a tunnel [47, 48] (for a picture see Ref. 32).

number of molecules forming the bilayer. This scaling is accounted for by the Helfrich–Servuss equations discussed in Sections IV and V.

Since then, several simulations appeared, confirming this shape of the bilayer isotherm [40, 45, 46] and also this pattern of size dependence [33, 45, 46]. Correlation with N, the number of molecules forming the bilayer, was also reported [33, 45, 46]. In particular, Ref. 46 gives plots very rich in data, unfortunately buried as an insert in a bigger figure—hence easily overlooked.

On the right-hand side of Fig. 6 we have made use of the information obtained [47, 48] in a simulation in which the bilayer was stretched so much that a hole was created [48]. This topological change, the hole, as a tunnel filled with solvent, has been seen before [37] but was not investigated. In that later work, Ref. 48, the lateral tension was determined also for those larger areas. It falls suddenly, as indicated in Fig. 6; the much lower tension of a bilayer with a hole does not change much when the nominal area $A_\perp = L^2$ is increased further [48]. The hole really is a tunnel filled with solvent [37]; the topology of the bilayer has changed. The thermodynamics and the work of creation of such tunnel holes has received attention recently on a large scale, no doubt in preparation for the next step of filling the tunnel by a swarm of foreign particles or one giant molecule—perhaps with modeling a protein in mind. We do not review this work because it is in full swing right now. Obviously, transport across a membrane is a most important topic.

But this tunnel-hole formation is another eccentricity of this sheet of a bilayer. It is surely due to the special setup of the intermolecular forces (cohesive energy) and to the constancy of N. An interface would simply pick up some molecules from either phase to make up its preferred local density; the bilayer does not do that because it does not exchange surfactant molecules with the surroundings.

Looking at the full bilayer isotherm and having become acquainted with the plateau in Γ, we may well ask what happened to the "equilibrium" tensionless state? It has essentially been submerged by the sudden change of slope or, rather, became more difficult to locate. As we can see from the plots, a locus of vanishing lateral tension is easy to find in small systems; with increase in size, its localization becomes somewhat ill-defined; the bilayer isotherm displays larger and larger curvature in the transition region while the transition region gets smaller and smaller and the plateau nearer and nearer $\Gamma = 0^-$. It is as if the free energy were to develop a cusp for truly large systems while the bilayer isotherm develops a sharp jump in the derivative: from null in the floppy plateau to K_l defined above. These notions have never been theoretically justified. We will show further discontinuities in other sections.

The state $\Gamma = 0$ is labeled as the equilibrium state, because of the relation with the free energy

$$\partial F / \partial A_\perp = \Gamma \tag{3}$$

When the bilayer isotherm is integrated over a [37], rather obviously a curve $F(A_\perp)$ is obtained with a minimum at a_0 [37]. It is F at constant numbers of particles, volume V, therefore total nominal density $\rho = \mathcal{N}/V$ and temperature. In this context, it appears that the tensionless state defined in terms of the bilayer isotherm is the correct equilibrium state with respect to variations of $A_\perp$ at the above conditions. The other tensionless state is discussed in Section IV.

In the tensionless state $\partial F/\partial A_\perp = 0$ the normal pressure is equal to the tangential pressure $P_{xx} = P_{yy} = P_{zz}$ so the system is "at rest." But physically it is an accident that in a deeper sense does not mean much. The bilayer is strongly anisotropic, and its elongated molecules are on the average parallel to the z-axis; stretching or compressing it along z is a process quite different from doing it along x or y; that the respective elements of the stress tensor should be equal is a coincidence. We searched for some understanding of the factors involved by splitting Γ into its partial contributions [40].

The general shape of a plateau at small areas and linear increase at large areas has also been seen in experiments on spherical vesicles [57] (see also Refs. 58–60). The authors also brought up [57] analogies with the transitions known from the field of polymers, thereby suggesting a transition in the first instance; we return to this unresolved topic in Section XI.

IV. STRUCTURE FACTOR

The structure factor $S(q)$ originates from the height–height correlation function $\langle h(x_1, y_1) h(x_2, y_2) \rangle$; in terms of the Fourier components we have

$$h_{\mathbf{q}} = \int_0^L \int_0^L dx\,dy \exp[i(q_x x + q_y y)] h(x, y) \tag{4}$$

$$S(q) = \langle h_{\mathbf{q}} h_{-\mathbf{q}} \rangle \tag{5}$$

This would be for the usual simulation box $L \times L \times L_z$. With circular averaging over Fourier components of $\mathbf{q} = (q_x, q_y)$, $S(q)$ depends on the scalar $q = |\mathbf{q}|$. The important point is that we have here a Fourier series: Because the position space is continuous, the Fourier space is discrete. The allowed values are

$$q_x = n(2\pi/L) \qquad (n \quad \text{integer}) \tag{6}$$

and similarly for q_y. Therefore q can never be smaller than $q_{00} = 2\pi/L$. The limit $q \to 0$ of the capillary wave theory is meaningful only after the limit $L \to \infty$ has been performed.

It is the capillary-wave theory that provides the first step in the interpretation of the structure factors of interfaces and/or bilayers. Embedded in a space of higher dimension, these objects can and do execute large fluctuations of shape by making excursions from the planar shape into the extra dimension now available. The fluctuations of a homogeneous phase are much milder and S is a small number related to compressibility, raising strongly only when q approaches the nearest-neighbor peak at (or near) $2\pi/\sigma$. σ is the collision diameter. In fact, this contribution is usually neglected.

The capillary-wave theory was formulated a very long time ago and has been rederived many times either in the context of the mechanics of fluids (see, e.g., Ref. 62) or in the context of the scattering experiments on liquid surfaces and interfaces. Its essence is very straightforward. The planar interface is flat, $h(x, y) = h_0$, and its area is $A = L^2$. The fluctuations produce an instantaneous shape $h(x, y)$ with an *increased* area $A = \sqrt{1 + |\nabla h|^2}$ which for small h and small gradients—that is, for small q—can be approximated by

$$A = L^2 + \sum_{\mathbf{q}} (1/2)q^2 h_{\mathbf{q}} h_{-\mathbf{q}} \tag{7}$$

The capillary-wave theory has been formulated for surfaces and interfaces; then routinely. the free energy increase is

$$\Delta F = \gamma \Delta A \tag{8}$$

where γ is the interfacial tension. In the usual fashion of the mesoscopic statistical theory, we call it a Hamiltonian, and the statistical sum over all available states (renamed to functional integral over $D[h]$) becomes the integral over the random variable h_q which can vary from $-\infty$ to $+\infty$. Now we have the Gaussian integrals

$$\int_{-\infty}^{+\infty} dx \exp(-\alpha x^2) = \sqrt{(\pi/\alpha)} \tag{9}$$

and

$$\int_{-\infty}^{+\infty} dx x^2 \exp(-\alpha x^2) = (1/2\alpha) \times \sqrt{(\pi/\alpha)} \tag{10}$$

With $\alpha = \beta\gamma = \gamma/kT$, the structure factor is obtained as

$$S(q) = kT/(\gamma q^2) \tag{11}$$

This is the "capillary wave divergence" resulting, in the last resort, from the Goldstone mode of our system [56]. In fact the shift of the whole thing along the z-axis by Δz does not change the free energy in the absence of external fields. It has been recognized that the external field is needed to localize the interface and also to make it planar. The best-known example is the gravitational field. Typically, a presence of an external field puts a constant in the denominator of Eq. (11), so that $S(q)$ is now $\sim 1/(D+\gamma q^2)$. Anticipating the importance of the curvature effects in bilayers, we add a q^4 term with κ the rigidity constant, so that S becomes

$$S(q) = kT/(D+\gamma q^2 + \kappa q^4) \tag{12}$$

For interfaces, κ does not matter; for bilayers and membranes, κq^4 is *the* main effect and γ can be omitted. This was the common and neat (thereby appealing) distinction prevalent in the literature of the 1980s and 1990s. For bilayers the external field practically is not needed. Then the divergent part of S should be

$$S(q) = kT/(\gamma q^2 + \kappa q^4) \tag{13}$$

In accordance with these views, the γ term is often neglected too so that the mesoscopic hamiltonian for a membrane or single bilayer becomes

$$H = \int \int dxdy(1/2)\kappa|\Delta h|^2 \tag{14}$$

or

$$H = \sum H_q, \qquad H_q = (\kappa/2)h_q h_{-q} q^4 \tag{15}$$

This is nothing else but a particular case of the celebrated Canham–Helfrich hamiltonian [55] (see Section V). Then $S(q)$ follows:

$$S(q) = \frac{kT}{\kappa q^4} \tag{16}$$

It was only in 1999 that such q^4 divergence was found in a simulation [35] for the very first time. But it was found only for one point on the bilayer isotherm (cf. Section III), namely for the tensionless state. For other states we therefore must come back to the more general form; that is, we have to introduce γ. But on reflection (not without it!) we do not know what γ, since we have two areas: the true area of the bilayer as it changes its shape (maybe the average of it) and the nominal area, the "projected area" equal to L^2. Thus we have arrived at the "two tensions issue" [32, 33, 37, 39], which deserves a particular discussion (see Section V).

The structure factor of the bilayer is determined in two ways. In earlier work, the x–y plane was divided into a rectangular grid of cells; the threadmill of Fourier analysis of equidistant data followed; then from instantaneous configurations the averages were obtained [12–18, 46]. Lately, however, h_q was computed by interpreting the coordinates of the particle, (x, y, z), as $(x, y, h(x, y))$— taking its z-coordinate as $h(x, y)$. The "particle" may be the head of the chain, or one can use the coordinates of the middle of the first bond in the chain—that is, of the a–b bond between the head and the next bead. In this way all gridding is avoided.

In all states along the bilayer isotherm, strong scattering at low q signals the capillary-wave divergence after Eqs. (11)–(13), or the κ-divergence after Eq. (16), or a mixture of both. With increasing q the strong scattering dies down and then $S(q)$ stays at low values to increase gradually up to the nearest-neighbor peak near $q_{n.n.} \simeq 2\pi/\sigma$. Further details depend on the location of the state point on the bilayer isotherm. We have identified (see Section III) two main regions with a transition region betweeen them. The "extended region," "EX," lies at high area per head, a, where the bilayer is extended, nearly flat. The "floppy region," "FL," lies at the other extreme of low a, where the bilayer is in the floppy state. In between lies the transition region.

In "EX" the capillary-wave $1/q^2$ divergence dominates. Just like for the lateral tension Γ, there is no visible size dependence. In fact, there is a weak shift in the region near the minimum of $S(q)$, apparently of no significance. At small $x = q^2$ the capillary-wave divergence $1/x$ ultimately overshadows any κ-contribution. Moving away from the strictly asymptotic region (i.e., to larger x), we find the second derivative $d^2 S/dx^2$ invariably positive and staying so through the minimum and up to the inflection, usually near $x \simeq 25$ or more ($q \simeq 5$ or more). Then the nearest-neighbor peak dominates from $x \simeq 25$ to $x \simeq 55$. Figure 7 illustrates these statements.

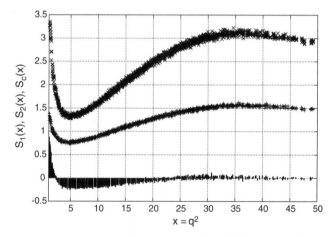

Figure 7. A typical structure factor $S(x)$ plotted against $x = q^2$ in an exceptionally wide range of x including the ubiquitous minimum and the nearest-neighbor peak near $x = 4\pi^2$. S_c is the cross-correlation between the monolayers (bottom of the figure), S_1 are for individuals monolayers, and S_2 is the sum of all three (top of the figure). The divergence at lowest q will be shown in other figures with logarithmic scale. These data are for the reverse dimers $N = 2880$ and $T = 1.9$ in the EX region, with $\Gamma = 0.47$ in LJ units as all our data are shown. Each S is a swarm of data points.

Incidentally, for smallest q the data per force become scarce—in the plot of Fig. 7 we cannot add any more points because all have been used: the first few (n_x, n_y) pairs in $\mathbf{q} = (2\pi/L)(n_x, n_y)$ are $(0, 1), (1, 0), (1, 1), (2, 0), (0, 2), \ldots$ and cannot be made denser other than by simulating afresh a larger system (i.e., with larger L).

The existence of a minimum—typically near $x \sim 2\text{–}9$, $q \sim 1\text{–}3$—suggests a sum of two components: one decreasing and the other increasing.

We cannot expect $S(q)$ to die out to nil when the small-q divergent part vanishes. The simulated system must have a bulk-like contribution to $S(q)$. However unimportant for the first few data points, it must show up for higher q. Some time ago we analyzed another case, that of a simulated planar liquid–vapor interface, and we found [63] that indeed the two-point correlation function $G(z_1, z_2; q)$ contained a sizable bulk-like part even in the asymptotic region. And Fig. 7, which is typical, shows that the raise toward the nearest-neighbor peak begins rather early. For these reasons we have fitted the data for $S(q)$ from simulations not to Eq. (13) but to an augmented expression

$$S_{\text{fit}} = \text{const}/(kx^2 + gx) + S_{\text{reg}}(x) \qquad (x \equiv q^2) \qquad (17)$$

with the regular part of S a purely empirical polynomial. A really satisfactory fit ought to capture the minimum near $q \sim 1\text{–}3$. The fitted k corresponds to the rigidity coefficient κ, and fitted g ought to correspond to a capillary-wave γ suggested by Eq. (13). Such fits have been reported [33, 39, 45] whereas in earlier work only Eq. (13) was used, not necessarily with least-square fitting [16–18] and not within any wider range of areas. The extended simulations over the entire bilayer isotherm are rare [37–39, 46].

Lowering a, we leave the "EX" region, but before reaching the "FL" region, we cross the tensionless state that displays the pure $1/q^4$ divergence [35]. Indeed a special place is due to the first empirical proof of the curvature q^4 divergence [35] at the tensionless state (or very close to it). The published plot shows most clearly the q^4 divergence. It also shows the common feature of all plots: positive second derivative $d^2 \log S/d \log x^2$. S appears to be drifting toward a minimum that unfortunately is beyond the data limit. The data were fit to the visible $1/q^4$ straight line in the logarithmic plot, but not quite so, because, in order to accommodate the deviation steadliy increasing with q, Eq. (16) was augmented to

$$S_{\text{GGL}}(q) \equiv \kappa/q^4 + \sigma_p/q^2 \qquad (18)$$

The added term was interpreted as the contribution of "protrusions" [32, 33, 35]. We have used for fitting $S(q)$ an expression with an extra $1/x$ term

$$S_p(x) = \text{const}/(kx^2 + gx) + p/x + S_{\text{reg}}(x), \qquad x \equiv q^2 \qquad (19)$$

as one of alternative expressions [37, 39], without a particularly notable success. But the importance and significance of Ref. 35 was that before this simulation was published, the $1/q^4$ divergence was a hypothetical theoretical prediction; since that date, the number of simulations reporting such divergences greatly increased to a dozen, starting from zero (see, e.g., Refs. 16–18, 37, 39, 45, and 46).

We have fitted a large part of our data [37–39, 61] to expression (15) and its modifications and the resulting parameters k and g were shown and discussed in some detail [39].

On further compression the bilayer transforms itself into the floppy state. In "FL," S of a floppy bilayer diverges not as $q \to 0^+$ but $S \to \infty$ as $q \to q^{\dagger}$. At $q = q^{\dagger}$ lies the asymptote (with $0 < q^{\dagger} < q_{00} = 2\pi/L$). This novel feature is due to that extraordinary behavior of a floppy bilayer we described in Section III and discussed in some detail [37]. Besides taking negative values, the lateral tension Γ becomes strongly dependent on the size [37, 39, 45]. If the area L^2 is sufficiently large, $|\Gamma|$ is very small in the floppy region. It is always so in all kinds of models [37–39, 45, 46]. Large L in this context means large N because $a = 2L^2/N$ must be small enough so that we are still in the FL floppy region. In Section V we discuss the theoretical background for such a behavior; here we examine what this implies for the structure factor $S(q)$. Fitting the data from the floppy region to Eq. (17) or to Eq. (19), one finds $g < 0$. This is to be expected since Γ is negative there. But then the denominator has a pole at $kx + g = 0$ (i.e., at $x^{\dagger} = -g/k$) or at $x^{\dagger} = -\Gamma/\kappa$ (which will be proven not accurate). As $x^{\dagger} > 0$, the statements of the capillary-wave theory have to be qualified; the divergence may be $1/q^2$ in general, but for negative g it is $1/(q^2 - x^{\dagger})$ [39] with amplitude $1/k$. The prefactor $(1/kx)$ often is numerically non-negligible.

The fitting of experimental data to preordained mathemtical expressions is fraught with well-known dangers only partly alleviated by the use of statistics developed for such purpose. It is very easy for an unsuspecting experimentalist to ask too much and to check not too much. With these dangers in view, one can develop a series of easy tests. First of all, looking at expressions (16)–(19), one sees that the quantity $xS(x)$ ought to behave like $1/(kx + g) + w_0 x + \cdots$ or like $1/(kx + g) + p + w_0 x + \cdots$. The importance of k (hopefully equal to κ) is more easily assessed. Then $S(x) \times x^2$ ought to behave like $x/(kx + g) + px + w_0 x^2 + \cdots$ (alternatively with $p = 0$). The function $x/(kx + g)$ is $\propto x$ for vanishing x and $1/k$ for large x; it is therefore a good test of data and of their representation. For the cases of negative g, the inverses are very useful. The data for $S(x)^{-1}/x$ ought to follow $kx + g$; that is, the data points ought to aim at a crossing of the x-axis at $x^{\dagger}$, instead of aiming at a crossing with y-axis. For a purely empirical fit, Pade approximants represent the simulation data extremely well, but the parameters mean nothing as far as one can see.

The expressions used, although derived from a well-known theory, were not fully satisfactory. For positive g, the fraction $1/(kx^2 + gx)$ always approaches $(1/gx)$ no matter what is the value of k, and sometimes this is not satisfactory. At the state with $g = 0$ exactly, the divergence is $1/q^4$. If we change the area of the bilayer by an infinitesimal amount, the ultimate divergence ought to jump to $1/q^2$. Yet $S(x)$ for states near that with $g = 0$ has a stronger divergence not well reproduced by the theoretical term—the fits are erratic and not too good: the expression ought to allow for a stronger divergence. There the purely empirical form $a/q^4 + b/q^2 + c$ does a much better job. With decreasing x, $S(x)$ goes up, often linearly on the log–log plot, but always with a positive second derivative— curving upward. But nonzero (i.e., positive) κ does the wrong thing: It increases the local slope but makes the second derivative *negative* which is never seen in the data. Except for the small and occasional hump in the middle range of x (see Section VI), all data show a positive second derivative on the log–log plot of S versus x.

Figure 8 ilustrates these problems by visualizing four functions: the compo- nents $1/x$ and $1/x^2$, their sum, and the standard fraction $1/(kx^2 + gx)$. The form predicted by the theory—with second derivative negative—is practically never seen in the data.

Figures 9 and 10 are to be compared with Fig. 8; these display two smooth functions representing exceptionally well real data, one for the EX and the other for the FL region. Two components $1/x$, $1/x^2$ are also shown.

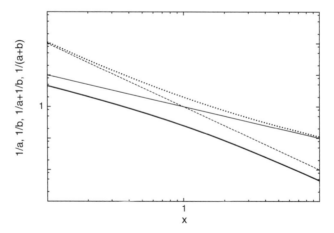

Figure 8. Graphical illustration of the components of the singular part of S with $x = q^2$ as variable: $a \propto x$ (capillary-wave divergence), full straight line; $b \propto x^2$ (curvature κ-divergence), dashed thin line; the sum of both fractions [Eq. (18)], thick dashes; and the fraction suggested by the extended capillary- wave theory $(a + b)^{-1}$, bold thick line with wrong curvature.

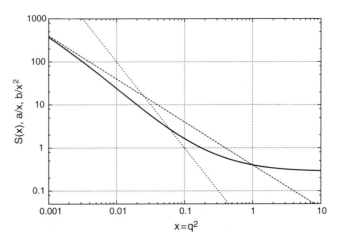

Figure 9. Graphical ilustration of the components of the singular part of S with $x = q^2$ as variable. Figure shows in logarithmic scale the function $s(x)$ that fitted the real data [chains with $\ell = 8$, big system with $N \simeq 4000$, $\mathcal{N} \simeq 104{,}000$, $T = 1$ (the EX region) and arbitrary components a/x, b/x^2].

We have to compare these plots with the series of structure factors along the bilayer isotherm, shown in Figs. 11–13, but before that we comment on Figs. 8–10. We can understand now the prevalent insistence of simulators on choosing the tensionless state: There we do not have to use the troublesome factor

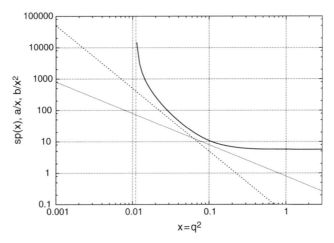

Figure 10. Now for the floppy region FL. Figure shows the function $sp(x)$ representing real data [chains with $\ell = 8$, big system with $N \simeq 4000$, $\mathcal{N} \simeq 121{,}000$, $T = 1$ (the FL region) and arbitrary components a/x, b/x^2]. Here only the small-gradient form $1/(kx^2 + gx)$ is able to produce the required asymptote (shown with vertical dashed line) at $x^\dagger = -g/k$.

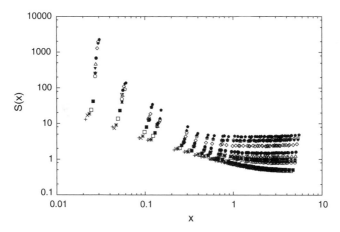

Figure 11. Series of structure factors along the bilayer isotherm. Data in this and the next two figures are for chain length $\ell = 4$ of the $N = 1800$ molecules in 49,000 solvent particles at $T = 1$ and high liquid-like density, averaged over directions of the q-vector and over all N—that is, taking the mean of two monolayers. $a \in (1.4, 2.1)$. Steeper curves belong to FL. Linear $1/x$ dependence can be detected in the EX region. $x = q^2$.

$(1/x)(1/(kx + g))$, but the theoretically expected pure κ-divergence $1/\kappa x^2$. The next step is to account for the (not very large) curving upward and there adding an empirical $1/q^2$ helps and therefore Eq. (18) can be used. The FL (floppy) region behavior (see Fig. 10) *can* be accounted for only by something like the theoretical term. A sum of inverse powers will not do.

Figure 11 shows data on series of structure factors $S(x)$ for several states along the bilayer isotherm for one particular bilayer. The log–log plot is needed to show the asymptotic region where S attains very large values. Each particular sign refers to one value of a, the area per head. We point out the smooth variation of shapes and values; the strongest divergence takes place in the FL floppy states.

The function $xS(x)$ (Fig. 12) proved to be very useful to accentuate the spread. The inverses $1/S$ and $1/(xS)$ are very useful to identify the floppy state of the bilayer and also to extract the parameters such as k and g. Some further data are discussed along with the normal–normal correlations in Section VI.

The plots in Figs. 11–13 displayed not quite the raw data, but the data averaged over the directions of the q-vector and also the mean (for each q) of S in the two monolayers. Figure 14 shows raw data (hence the scatter) of the inverse $1/xS$ for just one state of a bilayer in the EX–FL transition region close to the tensionless state: Its Γ is a very small negative number and g is positive $g > 0$. On the scale of the plot, the data aim at $(0, 0)$. The small-gradient small-q prediction is $kx + g \simeq kx$, and the purpose of Fig. 14 is to show how small the asymptotic region is. The arbitrary line $y = \text{const} \times x$ shoots up from $(0, 0)$ and becomes totally

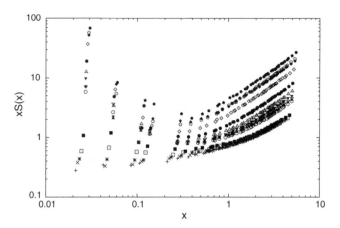

Figure 12. Data of Fig. 11 transformed to $xS(x)$; spreading more widely and removing the pole at $x = 0$ from Eq. (13) helps with fitting and checking fits. See text.

irrelevant for all data points to the right of the dotted vertical line indicating approximately where the maximum of this plot may lie. Thus it can be seen how small the asymptotic region is.

In our own work, three conclusions from the fits of $S(q)$ resulted. First, the fitted k was never negative, despite all the uncertainties and scatter. This is worth a remark, because for interfaces such fits to the corresponding $S(q)$ invariably produce a negative κ. It is obtained negative not only from fits but also from exact

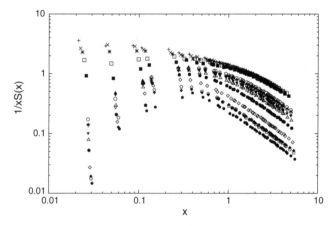

Figure 13. Data of Fig. 11 transformed to $[xS(x)]^{-1}$; any divergence is changed into a zero and the form $kx + g$ as $x \to 0$ is the easiest. The transition EX–FL to floppy, the new tensionless state with $g = 0$ (and $\Gamma \neq 0$), and the FL asymptote $x^{\dagger}$ are best localized from such plots. See text.

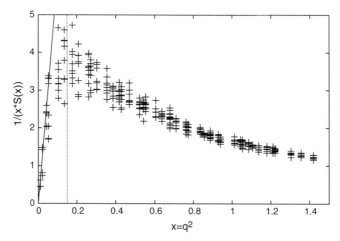

Figure 14. Raw data without any averaging, plotted as in Fig. 11 for just one state of the bilayer. This EX state is very close to both tensionless states. The figure is to show how small the asymptotic region is. The data can be interpreted as $kx + g$ at best up to the dotted vertical line. The straight line is to guide the eye.

calculation for a simple model [64]. As a result, augmenting the capillary-wave hamiltonian *for interfaces* by adding a q^4 rigidity-curvature term [i.e., from Eqs. (11) to (13)] is a dubious procedure for interfaces because a meaningful κ ought to be positive. For membranes, it works and the robustness of the positivity of fitted $k \simeq \kappa$ proves once again that the domination of curvature terms in membranes [55] is true.

Second, despite the scatter, the fitted parameters show a different dependence on the area in the floppy region FL from the dependence on the area in the extended region EX. This was already seen in Section III devoted to Γ and is also found in Sections VI, VII and VIII.

Third, the fitted g, from $S(q)$, is systematically higher than Γ [37, 39]. This finding gave rise to speculations about what became to be known among the initiated as "the two-tension" issue. My finding was later confirmed when "more careful and high-accuracy" simulations were done [29, 32, 33]. The difference between Γ and g was not noticed in an otherwise very thorough study [46]. Indeed the difference is not large but is firmly systematic. A discussion and some calculation can be found in the Section V.

Also, when one has seen such pictures as in Figs. 11–14, it becomes clear if it was not already that the "tensionless state" is just one point on the bilayer isotherm. The boiling point of a liquid is just one point on the vapor pressure curve, and its significance is tied to everyday experience under the atmospheric pressure. Close to it lies the triple point, and *this* point *has* physical significance. For a bilayer we have three points: the tensionless state after Γ and externally controlled $A_\perp = L^2$; the tensionless state after $g = 0$ (i.e., with pure κ-divergence of S); and the transition point between FL and EX. All three may be very close (in large systems) or not (in small systems). Clearly, when the analogy with a big spherical vesicle at rest on the laboratory table is brought in, $\Gamma = 0$ is the choice.

Whereas I have fitted data of $S(q)$ in the later work, the existence of two numerically different tensions was found by a different route. Namely, the authors were adjusting the simulation parameters in order to find the system (bilayer plus solvent) in that other tensionless state—that is, when the structure factor would show the purest κ-divergence $1/q^4$. In the course of these preparatory simulations, it was found that the desired tensionless state was shifted with respect to the expected tensionless state with $\Gamma = 0$ [29, 32, 33].

As a minor remark, we have not found any confirmation of the supposed $1/q^2$ dependence of a term representing the effect of protrusions [35]. One may well hypothesize that single molecules and groups of molecules of various sizes "protrude" out into the neighboring solvent down to the size of a single head "sticking out"; somewhat similar ideas were behind "the hat model" [65] (although for protrusions of larger sizes). Snapshots of a simulated bilayer, such as shown in many references, allow for such interpretations but have not been examined in any systematic manner. When a sufficiently large interval of q is scanned and the ubiquitous minimum appears on the plot of $S(q)$ versus q^2, then it becomes obvious that the plot is not a sum of two straight lines and the authors do not attempt to introduce an extra $1/q^2$ term into the equations [42].

The q-dependence in the form $1/q^2$ was obtained in quite another problem, namely for the repulsive interaction betwee *two* bilayers separated by a layer of water. Incidentally, such arrangements have been sucessfully described by bor-rowing from the theory of smectics. The repulsive interaction is strongly affected by the protrusions in two monolayers facing each other across the solvent. In a single bilayer the protrusions might well show as an increase of amplitude of undulations of short wavelengths, of the order of the size of the hypothetical group protruding. This may be an explanation of otherwise unexplained small humps in plots of $S(x)$ in the range $0.5 < x < 3$ appearing over the fitted line of $S_{fit}(x)$. The range of q is the $(0.7, 1.7)$ or wavelengths 3.6 to 9 collision diameters or rather half of that, 1.8 to 4.5 collision diameters. These numbers would square with a size of a few molecule heads. But the increased scattering is small and the interpretation is tenuous. Some small humps in the same region of q are seen in correlation of normals (see Section VI).

It is not easy to identify the earliest simulations that determined the structure factor. Certainly, it only became possible with increased computer power that allowed for longer simulation time large enough for undulations to develop. The 1999 paper [35] we quoted above was based on a 1997 paper [34] that has no mention of the structure factor. Two very important papers [16, 17] and [18] published in 2000 and 2001, are probably the very first along with Ref. 35. But the undulation modes were expected earlier (see, e.g., Ref. 1). An elaborate model with about 50 atoms per lipid with $N = 64, 256, 1024$, and 23 point-charge "waters" per one lipid was run [16] for about 30,000 hours parallel computer time. The runs were 10–60 ns long with a timestep 2 fs (i.e., 5–30 million timesteps), not counting preliminary equilibrations. But the results were worth it; one estimates the scatter of data points for $S(q)$ as roughly comparable to that obtained in simplified models [35, 37, 45], if not slightly better. Constant pressure ensemble was used, and scaling in x- and y-directions kept the system at the tensionless point at 1 atm and 20°C. The agreement of specific parameters with experimental data from the real world [3, 60] was noted. The structure factor was fitted to (13) for $q < q_0$ and to a $1/q^2$ dependence for $q > q_0$ with an arbitrary q_0. The total range appears to extend to something like 10–$20q_{00}$. At that time a determination of S with a relatively realistic model, confirming the form (13) and determining κ along with other parameters—not only of the right order of magnitude but also with good numerical agreement—was a great success. But the division at $q = q_0 = 1.5$ nm^{-1} and the introduction of $1/q^2$ term cannot be tested because the range of q's would have to be much larger. Although is it outside the scope of this review limited to simulations of simplified models, we cannot avoid refering to these important papers [16–18]. The other success of that work [16, 17], rarely repeated [18], was the determination of the properties of the peristaltic motion. This self-explanatory term refers to thickness fluctuations of the entire bilayer, in which the monolayers move in phase, so that the total thickness oscillates. The q-dependence is now $1/(k_1 q^4 + k_2)$, and the interplay of parameters was such that the crossover to constant value of $1/k_2$ occurred within the data range (Fig. 4b of Ref. 17). Also, the split of Γ into components that add up to produce the observed lateral tension [17] preceded by many years a similar, more restricted split [40] of Γ of simplest molecular model.

Seeing the considerable investment for that kind of work, one can understand the motivation for using simple, even very simplified, models.

V. SMALL-GRADIENT THEORY

A. Application of the Standard Capillary-Wave Theory

The classical reference [66] to the theoretical treatment of the structure factor calculated the areas of fluctuating nearly planar membrane and the amplitude of undulations. It was soon extended and amplified by the Lecture at Les Houches School for Theoretical Physics [67]. But all treatment in there begins with the Canham–Helfrich hamiltonian [55], the free energy density.

Less generally, in order to treat shape fluctuations of nearly planar membranes (bilayers), one introduces first of all the function $h(x, y)$ that gives the position

Figure 15.   The fluctuating membrane (thick line) and the projected area (thin line).

$z = h$ above (or below) the (x, y) plane of the coordinate system. In this way we presume it is a single-valued function—the shape is still not too far from planar. This kind of imagery is illustrated in Fig. 15.

Figure 15 shows a membrane with one end nailed to the origin and the other end fluctuating at will, perhaps like a flag flopping in the wind. Such an image underlies the idea of a persistence length, after which length, it is said, the membrane must have forgotten its initial orientation. This idea, apparently perfectly reasonable, was thought to be useful but, on the contrary, stimulated confusion. In simulations we use periodic boundary conditions (if not specified otherwise—almost never the case); in Fig. 15 then the line representing the membrane ought to come down to $z = 0$ at the far end of the plot. Even so, a theoretician might very well argue that I have only introduced a factor of 2 because in the middle . . . . Luckily, the present-day computer resources do not make possible simulations of boxes $L \times L \times L_z$, with L much larger than of the order of 10^2 (at liquid-like densities of the matter inside), and this quite precludes any relevance of persistence length.

Figure 15 makes apparent the difference between the true area of the bilayer (or membrane), $\bar{A}$, and its projection L^2 onto the area of the simulation box (or any coordinate system that might be chosen)—"the projected area." The difference or relation between the two areas was the concern of the quoted references [66, 67]. The function $h(x, y)$ on the square $L \times L$ is developed in the Fourier series, Eq. (4). The coefficients h_q are amplitudes of "undulations" of the membrane. Given the continuous function $h(x, y)$, its area is given by an exact formula of the differential geometry

$$A = \int \int dx dy \sqrt{1 + (dh/dx)^2 + (dh/dy)^2} \qquad (20)$$

But this is quite untractable, no published work can be found where this would be used; all we can do analytically is to introduce the ubiquitous small-gradient approximation. The square root is then $\sqrt{1 + X} = 1 + X/2 - \cdots + \cdots$ and with the Fourier representation (4) we have

$$A = L^2 + \sum_q (1/2) q^2 h_q h_{-q} \qquad (21)$$

So A is the instantaneous area in terms of the Fourier amplitudes of the instantaneous configuration $h(x, y)$. To obtain the thermal average

$$\langle A \rangle = \sum Ae^{-\beta H}/Z[H] \tag{22}$$

we need the hamitonian H. The Canham–Helfrich hamiltonian [55], which involves only curvatures, is suitably simplified and supplemented [67] with a "free energy related to $A-L^2$"—that is, with a surface-tension-like term $\gamma \Delta A$. In our notation

$$H = \int \int dxdy(1/2)(\kappa \nabla^2 h)^2 + (1/2)\gamma(A-L^2). \tag{23}$$

In Fourier representation.

$$H = \sum_q h_q h_{-q} \Psi(q) \tag{24}$$

where

$$\Psi(q) = (\kappa q^4 + \gamma q^2)/2. \tag{25}$$

Since $h_{-q} = c.c.(h_q)$, then

$$h_{-q}h_q = |h_q|^2. \tag{26}$$

In (24) the sum is over the Fourier vector $\mathbf{q}$; in (22) the amplitude h_q is the random variable to be integrated over:

$$\sum_{\text{states}} = \int D[h] = \prod_q \int_{-\infty}^{+\infty} dh_q \tag{27}$$

The essential simplification is that the quadratic (in h_q) Hamiltonian puts us in the realm of the Gaussian (free-field) theory. In our simple calculation, referring to (24), we have

$$\langle |h_q|^2 \rangle = kT/2\Psi(q) \tag{28}$$

This is most often called the "equipartition theorem" (half kT per one mode), which is valid only for Gaussian distributions—that is, for quadratic Hamiltonians. This is how the average area is calculated; Eqs. (21)–(23) give us

$$\langle A \rangle = L^2 + (1/2)^2 \sum_q kTq^2/\Psi(q) \tag{29}$$

or [66, 67]

$$\langle A \rangle = L^2 + (1/2) \sum_q kTq^2/(\kappa q^4 + \gamma q^2) \tag{30}$$

or

$$\langle A \rangle = L^2 + (1/2) \sum_q kT/(\kappa q^2 + \gamma) \tag{31}$$

Since we are after an estimate, we can replace the sum by an integral:

$$\sum_q (\ldots) \to (L/(2\pi))^2 \int 2\pi q \, dq (\ldots) \tag{32}$$

The integral is taken from the lowest $q_{00} = 2\pi/L$ to the upper cutoff, which is usually $q_{mx} = 2\pi/\sigma$ with σ the collision diameter of molecules or any such microscopic length.

The latter choice may be easily discounted because at the onset we have taken approximations valid only for long wavelengths and small gradients; therefore a choice of $q_{mx} = 2\pi/w$, where w is of the order of the width of the membrane, is much better justified. Also, one can improve the result in the sense of better numerical accuracy, by performing the sums without approximating the sums by integrals [16, 17, 45]; this is probably inconsistent.

But all we want is a gross estimate, so we proceed and find

$$(\langle A \rangle - L^2)/L^2 = (1/(8\pi\beta\kappa)) \log \left[\frac{\beta\kappa q_{mx}^2 + \gamma}{\beta\kappa q_{00}^2 + \gamma} \right] \tag{33}$$

This gives the membrane area "absorbed by the undulations." Originally, γ was identified as Γ, the lateral tension [66, 67]. At the tensionless state, (33) reduces to

$$\Delta A/A = kT/(4\pi\kappa) \log(L/\sigma) \tag{34}$$

Also, (33) has been simplified with approximations, and the experiments on vesicles (by pipette aspiration) have been fitted to simplified versions [57]. Next, one allows [66, 67] some stretching elasticity added in the form

$$(\Delta A/A)_{\text{elas}} = (1/v)\Gamma \tag{35}$$

so that (33) is supplemented with this term

$$(\Delta A/A) = (\Delta A/A)_{\text{und}} + (\Delta A/A)_{\text{elas}} \tag{36}$$

Everywhere A is the "projected area," $A_\perp = L^2$ in our notation.

Equations (35) and (36) are the really weak point of that theory. There is no doubt that any membrane, bilayer, or vesicle must have some stretching elasticity, but it must be coupled to the *true* membrane area, not to its projection (see Fig. 15). And proportionality to lateral tension Γ in (35) is not correct.

Independently of any theory of undulations and area absorption caused by them, one can consider the function $\Gamma(a)$ in the small neighborhood of the tensionless state. This was referred to as "elasticity equation"—the response of lateral tension to small change of area. It is usually written in the form

$$\Gamma(a) = \Gamma(a_0) + K \frac{(a - a_0)}{a_0} + \cdots \tag{37}$$

and the higher-order terms are neglected. With a_0 being the area per head in the tensionless state, we have $\Gamma(a_0) = 0$. Or in integrated form, having defined the respective free energy

$$\partial f / \partial a = 2\Gamma \tag{38}$$

$$f(a) = f(a_0) + K' \frac{(a - a_0)}{a_0} \tag{39}$$

The determination of K from experiments or simulations has been the subject of some debate [14, 18, 42]. The debate concerned the interpretation. But it is clear now that there are two "elasticity" coefficients: One, K, is tied to the projected area, and the other is tied to the true area of the membrane. Thus to be in the right, the coefficient $1/v$ of (35) and Ref. 68 ought to be tied to changes of the true, intrinsic area, and not to Γ but to the true intrinsic elastic tension.

We found that the function $\Gamma(a)$ in the floppy region is strongly size-dependent (Section III). Asking for the numerical value of K is just asking for the derivative at zero: slope at crossing of the ordinate a. For a small system this is easy, but for large systems where the tensionless point lies in the transition region of strong curvature, the slope may be difficult to determine. The limit from above may be different from the limit from below if the scale is not fine enough. And the resulting value will depend on size. Plots of simulation results [37, 39, 45, 46], especially in Ref. 45, testify to that. We comment in more detail on this in Section III, which is devoted to lateral tension.

In view of the size-independence of $\Gamma(a)$ in the EX extended region, it is tempting to define K_{ext} in terms of the slope extrapolated from the right—that is, from higher a to lower. This may be more meaningful physically than the size-dependent K at the true zero of $\Gamma(a)$.

When calculating the average area $\langle A \rangle$, the average amplitude was also found [67] as

$$\langle |h_{\mathbf{q}}|^2 \rangle = 2kT / (\kappa q^4 + \gamma q^2) \tag{40}$$

where γ is always identified with the lateral tension Γ.

This approach and the results given at Les Houches [67] form the basis of common and general knowledge and are used for the interpretation of experiments.

Apparently no distiction was then made between the extended bilayer and the compressed bilayer in the floppy state, although the term "floppy" was used and it is clear from the context that the distinction was recognized. The notion of "floppiness" came about from experiments on vesicles [57], and the existence of some kind of transition between was already suggested in 1990, but without apparent consequences for the simulation study of bilayers that were invariably investigated at the tensionless state.

Thus one finds in 1988 [67, page 225] a statement that "The crossover from the undulating but planar state to floppiness is preceded by ..." and then "any theory of the floppy state has eventually to include the self-interaction of the membrane" [67].

We have seen in Section III that in simulations the lateral tension reveals, when taken along the entire bilayer isotherm, the existence of a floppy region as distinct from the extended region. Apparently the existence of these two states or regions of bilayers was known but not much discussed. To this day, we do not have a theory of the transition EX–FL (see Section XI).

The early theory captures some essentials, but it ought to be improved.

B.   The Constraint of Membrane Intrinsic Area

It was much later, in the year 2001, that a conceptually important calculation was published as a brief Letter [68] in which the constraint of a fixed-membrane intrinsic (proper, true) area was discussed and introduced.

In the capillary wave theory for interfaces, as in the previous subsection, the flat interface is treated as given and the instantaneous *larger* area results as produced by fluctuations. In membranes and bilayers the roles are reversed: The membrane area is given and its projected area results, varying with fluctuations of the shape of the membrane if the external conditions permit. Such a constraint of given membrane area produces a situation familiar from the context of polymers; a single polymer is pictured as a chain of *fixed length* which can wiggle in various ways. Similarly, a bilayer has a fixed number of amphiphiles, N, which approximately translates into a fixed area $\bar{A}$—of course "approximately," because some limited stretching elasticity will have to be allowed. For now and for the sake of simplicity, we provisionally take the true membrane area to be strictly constant [68].

N stays constant during a simulation because the solubility of the (lipid) amphiphiles in the solvent is very low. In simulations, one very rarely sees an amphiphile dissolved in the solvent and practically no amphiphile ever leaves the bilayer.

Therefore unmodified application of the capillary-wave theory is not quite correct; the undulations of surfaces and interfaces are those of *open systems* exchanging particles with the surroundings—that is, with the coexisting phases. Indeed the best calculations are performed in the grand canonical ensemble [56]. A telling example is the surface of a perfect crystal: It changes shape *only* thanks to exchange of molecules or atoms with vapor.

The bilayer does not do that. On the contrary, the bilayer changes shape *only* by displacements of its constituent molecules. This was recognized in Ref. 68, but not always.

An introduction of the constraint turns out to be helpful in the resolution of the "two-tension issue." It is an important difficulty [33, 68, 69] which caused much confusion. Thanks to simulations [29, 33, 37, 39], one can resolve it.

The simulations [37, 39] have shown that the "surface tension" g determined from the structure factor is different from, always slightly larger than, the lateral tension Γ. Since Γ is coupled to the projected area $A_\perp$, it is natural to hypothesize that $g \approx \gamma$ is the tension coupled to the true area of the interface. This view was taken in Refs. 29 and 33; the original and quick calculation [33] will be described too.

Thus we must construct the theory of undulations afresh, taking this constraint into account. We then calculate the height–height correlation function under the constraint and demonstrate how a new tension results, always a bit larger than the lateral tension.

The important letter [68] we mentioned starts from the membrane microscopic hamiltonian (14) and proceeds to construct the partition function Z with the constraint introduced as a Dirac delta function. But we modify it below because it introduces *three* areas, with micropipette aspiration experiments in view. In such experiments the projected area L^2 stays constant, the proper area of the membrane $\bar{A}$ is also constant by hypothesis, and the observed (coarse-grained) area results from pulling the membrane mechanically at both $L, \bar{A} = \text{const}$. We do not have such a situation in simulations of bilayers and therefore we do not introduce the third area. We use all the approximations of small q and approximate all calculations of the areas by (32) or equivalent.

Thus the constancy of membrane true area $\bar{A}$ requires

$$A[h] = L^2 + \sum_q (1/2)q^2 h_q h_{-q} = \bar{A} \tag{41}$$

The microscopic Hamiltonian was taken as in (14), or

$$H_c = \sum_{\mathbf{q}} (\kappa/2) X q^4 \tag{42}$$

The unconstrained partition function is

$$Z = \int D[h]\exp[-\beta H_c] \tag{43}$$

$$Z = \prod_q \int_{-\infty}^{+\infty} dh_q \exp[-\beta \kappa X/2] \tag{44}$$

and the constrained one

$$Z = \prod_q \int dh_q \int dk \exp(E) \tag{45}$$

$$E \equiv ik(\bar{A} - L^2) - \sum_q X(ikq^2/2 + \beta\kappa/2) \tag{46}$$

To alleviate notation we abbreviate

$$h_q h_{-q} \to X$$

The Dirac delta function ensures that $\bar{A} - A[h] = 0$ at all times. Z has not been evaluated either numerically or analytically, but as an approximation it could be evaluated at the saddle point $\lambda^* \equiv ik^*$; Z is then a Gaussian integral thanks to the low-q approximations

$$\log Z = \lambda^*(\bar{A} - L^2) - \sum_q (1/2)\log(\beta\kappa q^4 + \lambda^* q^2) \tag{47}$$

omitting π and numerical factors, which will cancel in all applications. The condition

$$d \log Z/d\lambda^* = 0 \tag{48}$$

determines the value of λ^* which is to be substituted into (47). Approximating the sum with the integral after (32), we recover the mathematical form of (33) in the equation for λ^*:

$$(\bar{A} - L^2)/L^2 = (1/(8\pi\beta\kappa))\log\left[\frac{\beta\kappa q_{mx}^2 + \lambda^*}{\beta\kappa q_{00}^2 + \lambda^*}\right] \tag{49}$$

The advantage of this equation (i.e., with integral done) is that it can be solved analytically for λ^*. Then $\log Z$ is also expressed analytically in terms of λ^* and of the integral

$$I \equiv \int \log(Cx^2 + Dx)\, dx \tag{50}$$

taken between $x = q_{00}^2$ and $x = q_{mx}^2$, with $C = \beta\kappa/2$ and $D = \lambda^*/2$. Here and in numerical work, we allow the q-values between q_{00} and $q_{mx} = q_w \equiv (2\pi/w)$, where w is the width of the membrane. The range $0 \le q < 2\pi/L$ is not included in the summations, and q higher than $\sim 2\pi/w$ contradicts the assumption of low q.

So far we have not departed far from the brief Letter [68], which also recovered the mathematical form of (49) first obtained in classical references [66, 67] as described above. This lack of essentially new result must be a consequence of the saddle-point approximation; nevertheless, it is a satisfactory result; it obtains from a well-defined model (Hamiltonian) and well-defined approximations. It may be a starting point for improvements.

We calculate now the *lateral tension* Γ, absent from the original Letter [68]. It is

$$\Gamma = -d\log Z/d(L^2) = -\lambda^* - I/(8\pi) - (\pi/2L^2)\log(Cq_{00}^4 + Dq_{00}^2) \qquad (51)$$

Figure 16 shows an example of calculated lateral tension against the projected area $A_\perp = L^2$ and the saddle point λ^*, for some arbitrary values of the parameters; their difference is apparent, but both quantities show a general shape that is already not too bad: either flat near respective zero, negative values for small areas (the "FL" floppy region), or sharp rise at higher areas with positive values (the extended "EX" region). Finer details as we know them from simulations, it has to be said, are difficult to reproduce with these equations. The divergence of both λ^* and

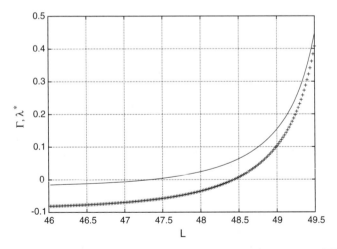

Figure 16. Plot of lateral tension Γ ((+)-signs) and of saddle point λ^* against $L = \sqrt{A_\perp}$ with fixed area $\bar{A} = 50^2$, $\beta\kappa = 1$, and $q_{mx} = 2\pi/11$. Above $L = 48.44$, both are positive; between 47.33 and 48.44 we have $\lambda^* > 0$ and $\Gamma < 0$; below 47.33 both are negative—we are firmly in the floppy region. See text.

Γ for areas L^2 approaching $\bar{A}$ (50^2 for this plot) is spurious and results from the strict equality of $A[h]$ and $\bar{A}$. When this is relaxed, the divergence disappears.

We calculate now—for the first time—the height–height correlation function, $S(q)$, under the constraint of constant true membrane area $\bar{A}$. Choosing a particular (allowed) value $q = Q$, we seek the average of $h_Q h_{-Q}$; for all other values of q the integrals over dh_q remain unchanged and we obtain $S(Q)$ as

$$\langle h_Q h_{-Q} \rangle = Z(\lambda^*)^{-1} \prod_{q \neq Q} \int dh_q \exp[E] \times \int dh_Q (h_Q h_{-Q}) \exp[E_Q] \qquad (52)$$

with

$$E \equiv \lambda(\bar{A} - L^2) - \sum_q X(\beta\kappa q^4 + \lambda q^2)/2 \qquad (53)$$

$$E_Q \equiv -h_Q h_{-Q}(\beta\kappa Q^4 + \lambda Q^2)/2 \qquad (54)$$

The constraint is applied twice: to $\int D[h]$ in the numerator and to $\int D[h]$ of the partition function in the denominator. There result two saddle points: the λ^* for Z and the λ^{**} for the numerator. Evaluating the Gaussian integrals and complementing the product by the missing Q-factor to form $Z(\lambda^{**})$, we obtain

$$S(Q) = \langle h_Q h_{-Q} \rangle = 1/(\beta\kappa Q^4 + \lambda^{**} Q^2) \times Z(\lambda^{**})/Z(\lambda^*) \qquad (55)$$

The new condition reads

$$(\bar{A} - L^2)/L^2 - (1/(8\pi\beta\kappa))\log\left[\frac{\beta\kappa q_{mx}^2 + \lambda}{\beta\kappa q_{00}^2 + \lambda}\right] - 1/(L^2(\beta\kappa Q^2 + \lambda)) = 0 \qquad (56)$$

The solution of it, λ^{**}, is used to calculate $Z(\lambda^{**})$ and $S(Q)$. The condition for λ^{**} is the left-hand side of the condition for λ^*, (49), plus the correction term. The correction term has a factor $1/L^2$ which makes it a finite-size correction, of that order. Unlike the condition (49) for λ^*, (56) is a transcendental equation for λ.

We have solved for λ^* and numerically for λ^{**} for three examples: for the extended state, with $A_\perp = L^2$ close to $\bar{A}$; then for the floppy state, deep inside, and then for an intermediate state near the transition to the floppy state. In all three cases the shift $\lambda^{**} - \lambda^*$ was positive. It is also apparent that it is a finite-size correction, of the order of $1/A_\perp = 1/L^2$. There are several highly satisfactory features of this result. The known expression—that is, the general form for $S(q)$— is recovered, but with a correction factor. The value of λ at the saddle point depends on Q; therefore the "effective γ" that will be seen when data on $S(q)$ will be fitted

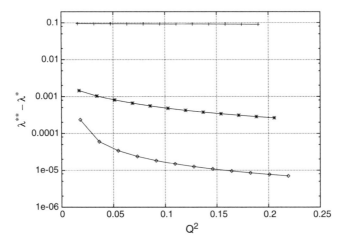

Figure 17. The difference $\lambda^* * -\lambda^*$ plotted against chosen Q^2 of $S(Q)$ for three chosen examples of L : 49.9 (the extended region), 48.0 (the ambiguous region), and 46.5 (the floppy region). See caption for Fig. 16. When compared with the values of λ^* in Fig. 16, these differences are seen to be small (smallest in the floppy region).

ought to depend on Q. This realizes the theoretician's dream about Q-dependent parameters, but it is not imposed, it results from a well-defined theory. Thus we found $\beta g = \lambda^{**}(Q)$.

We show a common plot for three points on the bilayer isotherm of the differences between the two saddle points (Fig. 17).

The difference is small—it is a finite-size effect—and it will be small for other values of the input parameters. It is invariably positive. Finally, we show in Fig. 18 the calculated structure factor $S(Q)$ plotted against $x = Q^2$ for the three chosen examples of different $L = \sqrt{A_\perp}$. Somewhere between the two might be found an ideally accurate $1/Q^4$ slope.

Neglecting the difference between λ^* and λ^{**}, we find

$$S(Q) = 1/(\beta\kappa Q^4 + \lambda^* Q^2) \tag{57}$$

This demonstrates the identification of the tension with the saddle point [68]; in this approximation $\beta\gamma \simeq \lambda^*$ results.

Whether we have arrived at the condition (49) for λ^* via the constraint calculation or via the capillary-wave theory extensions, both equations (49) (for λ) and (51) (for Γ) produce plots qualitatively resembling the "experimental" bilayer isotherms. See Figs. 1–6 in Section III. Already on the basis of the estimates [66, 67], we can come closer to explaining the peculiar behavior of the lateral tension Γ in the floppy region. Γ there is negative, varying little with area $A_\perp = L^2$. Its small

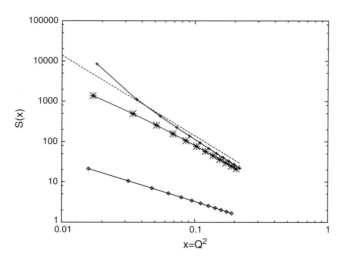

Figure 18. Plot of calculated $S(x)$ according to Eq. (55) against $x = Q^2$, for the same three examples as shown in Fig. 17. S for the case of $L = 49.9$ (extended region) follows quasi-exactly the capillary wave divergence const/x whereas for the other two cases the κ-divergence is present. For the parameters see caption to Fig. 16. The dotted line indicates the slope of pure $1/Q^4$ divergence.

absolute values vanish as $1/N$ at constant specific area a. a is the area per head, $a = 2L^2$; N, N is the number of heads, equal to the number of molecules making up the bilayer. The vanishing of Γ is understood from (33) as follows. On compression the left-hand side takes large values and on the right-hand side then the denominator under the sign of the logarithm must grow exponentially; hence $(2\pi/L)^2 + \Gamma \sim \epsilon$, where $\epsilon > 0$ is a small number. Thus Γ is negative and we have $|\Gamma| \sim L^{-2}$ for fixed N. Determining ϵ is beyond the accuracy of any computer simulation; it is neglected without any visible effect. Alternatively, comparing systems of different sizes at the same specific area a, we have $|\Gamma| \sim N^{-1}$ for fixed a. Thus (33) satisfactorily accounts for that very unusual feature of the bilayer isotherm. We discuss why it happens, in simple physical terms, in Section XI.

The constraint of strict equality of the instantaneous area to $\bar{A}$ can be relaxed [68]. Let us assume that the membrane area is not strictly constant but is distributed with some probability function sharply peaked about $\bar{A}$—for example, a bell-shaped Gaussian; then the Dirac delta function is replaced by such a finite width distribution

$$\delta(A[h] - \bar{A}) \to \text{const} \exp[-v(A[h] - \bar{A})] \tag{58}$$

This can be incorporated into the Hamiltonian. An equivalent formulation was given in the quoted Letter [68]. But again, with inevitable saddle-point approximation, the final working equations were not different from those proposed earlier heuristically [66, 67].

The work that confirmed and extended our finding of two tensions [33] contained an interesting interpretation as follows. The fluctuating area (21), on averaging, obeys (29)–(31) where the summation is over all q-vectors. One introduces now the quantity N', the "number of membrane patches which fluctuate independently." But the summation in (21) or (31) is over independent wave-modes, so that

$$\sum_q 1 = N' \tag{59}$$

In the large L limit, N' can be estimated by replacing the sum by an integral—that is, using (32) so that

$$N' \simeq L^2(q_{mx}^2 - q_{00}^2)/4\pi \tag{60}$$

or with bilayer width near 6.3 $q_{mx} = 1$ and $q_{00} \to 0$ for a large system,

$$N' \simeq L^2/4\pi = A_\perp/4\pi \tag{61}$$

Having assumed at the onset the Hamiltonian

$$H = \gamma A + H_c \tag{62}$$

with H_c the same as (30)–(31), one calculates the lateral tension Γ from the free energy calculated in the Gaussian approximation:

$$\Gamma = \gamma + (kT/(2L^2)) \sum_q (\gamma/(\kappa q^2 + \gamma) - 1) \tag{63}$$

Introducing N':

$$\Gamma A_\perp = \gamma\langle A\rangle - kTN'/2 \tag{64}$$

We found Eq. (23) of the cited paper [33].

It was assumed from the onset that there is a γ in the Hamiltonian coupled to the true membrane area A as a ΓA term. $\bar{A}$ was not assumed constant.

We can readily follow this and allow a nonzero γ coupled to the membrane area. We note that a form γA is used [40, 46] and the older form $K(A - A_0)/A_0)^2$ is abandoned. When γA is introduced into the microscopic Hamiltonian, the entire calculation is unchanged with everywhere λ replaced by $\lambda + \gamma/kT$. It could not be otherwise with fixed value of $\bar{A}$. Also, the condition for λ^{**} is now a condition for $\lambda^{**} + \gamma$.

By direct differentiation of the Gaussian partition function (44), one obtains the same final result, but then the interesting idea of N' and the quick estimates [33] are lost.

Another idea [70] was to consider a flat membrane in a rectangular box and apply a virtual deformation just like one does in the elastic theory of perfect crystals. The resulting expression for the tension in terms of the averages of microstates can also be written [70] in compact form:

$$\sigma_{FP} = L_z(C_{xzxz} + C_{yzyz} - P_{xz} - P_{yz}) \tag{65}$$

where C's are ordinary elastic constants and P's are elements of the stress tensor. This new tension was compared with the lateral tension Γ, and the difference was said to be equivalent to a certain shear modulus. The latter ought to vanish "in the termodynamic limit", and therefore the new tension σ_{FP} ought to become equal to the lateral tension Γ. This is plausible no doubt as the theory compares *two flat* surfaces.

Finally, the latest attempt [69] at the "two tensions issue" testifies to the growing awareness of these unresolved problems. As the authors point out, "when membrane tension is referred to, it is not always clear whether σ, τ, or r is meant" (γ, Γ, *or* g in our notation). Conceptual errors in previous theoretical work are pointed out. Via a continuum-elastic theory, the authors arrive at the same equations in terms of Fourier amplitudes [33] discussed above.

We turn now to the "toy model" of a single mode $h(x, y) = h_m \cos(qx)$ which already has produced [71] instructive results for the difficult quantity of the correct area

$$A \equiv \int dx dy \sqrt{1 + |\nabla h|^2}$$

For the single mode, A was calculated as a complete elliptic integral of the second kind. Here we can also do the calculation. We have in general

$$Z = \prod_q \int dh_q \int dk \exp(E) \tag{66}$$

and if h is made of only two modes h_q and h_{-q}, we obtain

$$Z_1 = \int_{-\infty}^{+\infty} dk \, 2e^{ik(\bar{A} - L^2)} / (\beta\kappa + \beta\gamma + q^2 ik) \tag{67}$$

The integral can be done analytically, with the result

$$Z_1 = \frac{(2\pi)}{q^2} e^{-(\bar{A} - L^2)\beta(\kappa q^2 + \gamma)} \tag{68}$$

This has a novel appearance, introduces an exponential dependence, and thus contains hints so far unexplored. Moreover, removing the constraint

$$\int_{L^2}^{\infty} d\bar{A}\, Z_1(\bar{A}) = \frac{2\pi}{\beta\kappa q^4 + \beta\kappa q^2} \tag{69}$$

reproduces the unconstrained Z.

VI. CORRELATION OF SURFACE NORMALS

Just like the height–height correlation, the angle–angle correlation is an interesting characteristic of a surface and its shape. In our case of a simulated membrane-like object, it is also a chacteristic of the fluctuations. If one starts with the assumption of a function $h(x, y)$ that gives the shape of the surface, then the normal, a unit vector, is defined as

$$\mathbf{N} = (-h_x, -h_y, +1)/s \tag{70}$$

with

$$s \equiv \sqrt{1 + h_x^2 + h_y^2} \tag{71}$$

In terms of angles we have $\sin\theta\cos\varphi$, $\sin\theta\sin\varphi$, and $\cos\theta$. The flat surface would have $\mathbf{N}_0 = (0, 0, 1)$ for a normal defined to point "upward"—that is, in the direction of increasing z-coordinate. Thus for a bilayer we define $\mathbf{N}_0 = (0, 0, 1)$ for the "upper" monolayer and $\mathbf{N}_0 = (0, 0, -1)$ for the "lower" monolayer. $\mathbf{N}$ varies from point to point, is $\mathbf{N}(x, y)$, and the correlation is the average of the scalar product of $\mathbf{N}$'s taken at two points. But to study fluctuations and deviation of the shape from ideal flatness, it is better to use $\delta\mathbf{N} = \mathbf{N} - \mathbf{N}_0$ so that

$$C = \langle \delta\mathbf{N} \cdot \delta\mathbf{N}' \rangle \tag{72}$$

where $\mathbf{N}$ is taken at point x, y, and $\mathbf{N}'$ at another point x', y'. Any function of position x, y inside the simulation box L by L can be expanded in Fourier series just like $h(x, y)$ was in Section V. Just like for the height–height correlation, because of the translational and rotational invariance we end up with a function of one vector $\mathbf{q} = (q_x, q_y) = \left(\frac{2\pi}{L} n_x, \frac{2\pi}{L} n_y\right)$ with n_x, n_y any integer. On averaging the results over the angle $\varphi \in (0, 2\pi)$, functions of one variable, the scalar $q = |\mathbf{q}|$, are obtained. The function $\tilde{C}(q)$ can be determined in a simulation. In our atomistic simulation we take the orientations of molecules for the definition of

$$\mathbf{N}(q) = (1/M_l) \sum_{j=1}^{M_l} \exp(i\mathbf{q} \cdot \mathbf{r}_j)\mathbf{N}_j \tag{73}$$

with M_l molecules in the monolayer l, $l = 1, 2$. And the normal $\mathbf{N}_j$ of the molecule j is defined as parallel to the first bond (a–b bond), just as it was in the previous section, except that the z-component is not $\cos\theta$ but $\cos\theta - 1$ for the upper monolayer and $\cos\theta + 1$ for the lower monolayer. In practical computation, one uses the angles θ, φ very simply determined. Because of properties of complex Fourier series the computation is of the order M_l and not M_l^2. Explicitly forming the scalar product

$$\tilde{C}(q) = \sum_{\alpha = x,y,z} \mathbf{N}_\alpha(\mathbf{q})\mathbf{N}_\alpha(-\mathbf{q}) \tag{74}$$

for each monolayer separately, we average over the simulation run:

$$\bar{C} = \langle \tilde{C}(q) \rangle \tag{75}$$

The results show many interesting features, above all a divergence for small q, but weaker than the height–height correlation, the structure factor S. The prediction of the small gradient theory is the following. Neglecting all the gradients, $\delta\mathbf{N}$ is $(-h_x, -h_y, 0)$ and therefore

$$\tilde{C} = q^2 h_{\mathbf{q}} h_{-\mathbf{q}} + \cdots \tag{76}$$

and averaging over the Boltzmann distribution with a Gaussian Hamiltonian 24, we obtain

$$\bar{C}_{\text{sm.gr.}} = \frac{xkT}{\kappa x^2 + \gamma x} \tag{77}$$

with $x \equiv q^2$. But we recognize $S(x)$ [Eqs. (13) and (28)] of the small-gradient theory, so there also follows a prediction:

$$\bar{C}_{\text{sm.gr.}}(x) = S_{\text{sm.gr.}}(x)/x \tag{78}$$

Of course, the small-gradient theory and its results can be valid at vanishing q, at best at q small enough.

In all our runs we have determined simultaneously all the various averages, and all runs were done along the bilayer isotherms for the model systems enumerated earlier: chain molecules $\ell = 8, 4, 1$ with the special feature of dimers $\ell = 1$ which were reverse dimers [37, 39]. Figure 19 shows a series of correlation of normals along the isotherm $T = 1.0$ for a series of a's at constant $(N, \mathcal{N})$, plotted against $x = q^2$. There are more than one value per one value of x for two reasons: The averages for the "upper" and "lower" monolayer may differ and the system area is a square.

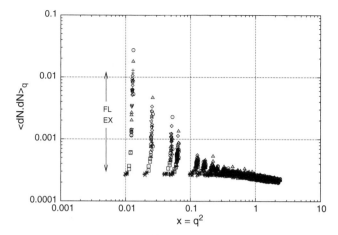

Figure 19. Normal–normal correlations $\bar{C}(q) \equiv \langle \delta\mathbf{N} \cdot \delta\mathbf{N} \rangle_q$ for a series of areas per head a along the bilayer isotherm, plotted against $x = q^2$. $T = 1, \ell = 4, N \simeq 4000$. The extended region "EX" of larger a's and larger Γ and the "FL" region of smaller a's and floppy bilayer are indicated by arrows. Arbitrary scale.

The apparent divergence of the form $1/(kx + g)$ results from several factors: the size of the region of reasonable accuracy of the square-gradient theory (this expression), say $0 \leq x < x_m^{sq}$, the position of the lowest data point $x_{00} = 4\pi^2/L^2$, and the size of the asymptotic region where other contributions to $\bar{C}$ do not matter. Interplay of these parameters determines the numbers produced and the actual look of the plot. Presence of positive g should flatten any curve $\bar{C}$ versus x compared to $1/kx$ (as $x \to 0$), but this could not be guessed by looking at Fig. 19; such flattening occurs well beyond the lowest data point.

In Fig. 19 the two regions, EX and FL, are indicated. The division was made according to the value of the lateral tension: positive or negative. The divergence appears milder in the EX region (see Section III). It gets stronger in the FL region where eventually a true divergence takes place. Just as for $S(x)$, in the case of negative g ($g \neq \Gamma$ but is close), $1/(kx + g)$ has a pole at $x^\dagger = -g/k > 0$ and becomes $(1/k)(1/(x - x^\dagger))$.

The data vary smoothly on the right-hand side of Fig. 19; $x = 1$ already is well outside the asymptotic region dominated by the term discussed above. It can be hardly seen in Fig. 19, but just as for the structure factor, there is a minimum at $x \sim 5 - 10$. A better exposition is obtained by examining the inverse, $1/\bar{C}$. As the small-gradient theory suggests, the *inverse* ought to behave linearly as $\kappa x + \gamma$.

Now the extrapolation to $x \to 0$ (or $1/\bar{C} \to 0$). Figure 20 shows a plot of $1/\bar{C}$ against $x = q^2$ for the EX part of the series. All bilayers had a positive Γ. To guide the eye, an arbitrary straight line was drawn; above it the data points curve upward and below it the data points curve downward. The linearity predicted by

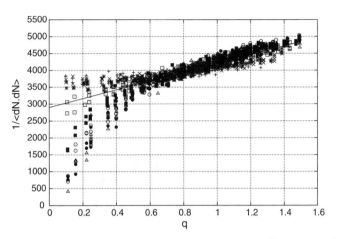

Figure 20.   The "EX" part of data shown in Fig. 19 plotted as inverses $1/\bar{C}$ against $x = q^2$. The line is arbitrary, all points had positive $\Gamma > 0$. Top monolayer only.

the square-gradient theory can hardly be confirmed—only for the four most-extended bilayers above the straight line. Then apparent κ in the guise of the empirical k is near zero. All less extended bilayers produce nonlinearity, as seen in the Fig. 20. In fact these series (e.g., the triangles) look quite like the data for the floppy region FL shown in Fig. 21. We can argue that for the values of k and g we must take only the first few data points since all the other ones already are

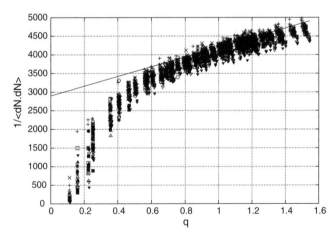

Figure 21.   The "FL" part of data shown in Fig. 19 plotted as in Fig. 20. The line is the same as in Fig. 20.

influenced by terms not accounted for; perhaps we should even take only the first two data points. As one can see at a glance, one then obtains a strongly varying k, $k(a)$, and the expected $g(a)$ similar to but not equal $\Gamma(a)$. We have done such detailed fitting in the past for the reverse dimers and published a few figures showing plots of these. Similar, though perhaps slightly smaller, scatter is produced with these data for chain-like molecules. The limit $g < 0$ is obtained (see Fig. 20) with the lowest data points, although only data series with $\Gamma > 0$ are included here.

The data for negative lateral tensions $\Gamma < 0$, thus for FL, the floppy region, are shown in an identical manner in Fig. 21. The arbitrary straight line from Fig. 20 is replotted here. Clearly all inverses of $\bar{C}(q)$ do not aim at the y-axis but point each to its own $x^\dagger$. Although the floppy bilayers show chaotic loose spatial structures (see also Sections III and Section XI), the parameters k and g are much more regular, k nearly constant, and inspire much better confidence. So it was with the reverse dimers [37].

Although from these three figures the transitions EX $\rightarrow$ FL and FL $\rightarrow$ EX appear smooth and structureless, a closer look reveals some interesting change of shape. We have chosen the smoothest example (i.e., with least scatter) of the value of $\bar{C}$ at the lowest data point—that is, at $x = 4\pi^2/L^2$ for each system along one bilayer isotherm.

Figure 22 shows both monolayers; $\bar{C}$ plotted against a is nearly constant in the EX region and starts to rise markedly in the floppy region, to flatten at

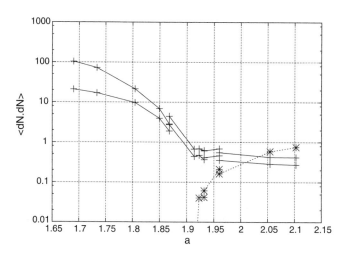

Figure 22. The data point of $\bar{C}$ for lowest q, $q = 2\pi/L$, plotted against area per head a for top and bottom monolayers (plus signs). $\Gamma(a) > 0$ is plotted (with star signs) to show the coincidence of breaks. $\ell = 8$, $T = 1$, $N \simeq 4000$.

the lowest data points (where the floppy bilayer is near disintegration due to its compression). For the sake of comparison, the lateral tension is plotted with stars and a dashed line. Clearly, the change of shape coincides with vanishing Γ or very close to it.

That $\langle \delta \mathbf{N} \cdot \delta \mathbf{N} \rangle$ at large distances (small q) should be small in the nearly flat extended bilayer but large in the floppy, chaotic, much more random, structure of the floppy bilayer may appear counterintuitive. But this time the square-gradient theory is on the side of the experiment: $1/(kq^2 + g)$ will be small for large positive g (which is near large positive Γ) and will be larger if g gets smaller and then negative. To satisfy the intuitive expectation, we would probably have to introduce conditional probabilities in the manner of $P(X_1, X_2)/P(X_1)$ or $(P(X_1, X_2) - P(X_1)P(X_2))/P(X_1)$ as Ciach has done for interfaces with great success [72].

Now we show the prediction (78) relating $\bar{C}$ to the structure factor S and conversely. We have chosen two plots, one for EX (Fig. 23) and one for FL (Fig. 24) and the procedure to divide $\bar{C}$ by x to bring it to equality with $S(x)$. S is shown with a line, and filled boxes represent $\langle \delta \mathbf{N} \cdot \delta \mathbf{N} \rangle / q^2$. The unchanged $\bar{C}$ is plotted for comparison with $(+)$-signs. The agreement is better for the floppy bilayer, and this is understandable: The latter is more like a liquid interface for which the equations of the standard capillary-wave theory are asymptotically valid. A general outlook is covered by the theory, though we have shown the best of all such plots.

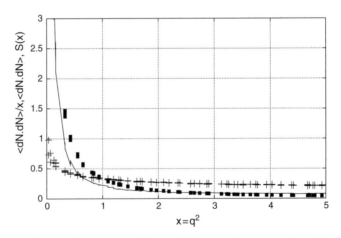

Figure 23. Check of the prediction (78) in the "EX" region: The structure factor S is plotted with a line, $\bar{C}/q^2$ is plotted with squares, $\bar{C}$ is plotted with plus signs—against $x = q^2$. The best such agreement is for the smaller system $N = 2880$ of reverse dimers ($T = 1.9$).

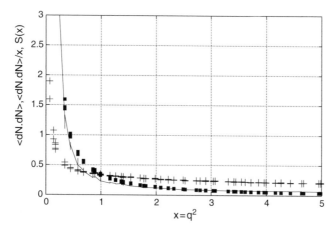

Figure 24. Check of the prediction (78) in the "FL" region. See caption to Fig. 23.

Finally, the results for the region of larger x are illustrated by the next two figures. Fig. 25 shows one $\bar{C}$ and its inverse plotted against x; the nearest-neighbor peak is seen near $x \approx (2\pi^2) \approx 40$. One very rarely, in fact never, collects the averages for q-vectors that large. It may happen more often that a simulator comes close to the the ubiquitous minimum (see, e.g., Ref. 45), here near $x \sim 10$ (i.e., $q \sim 3$). In order to show well the first few points belonging to the asymptotic region, the scale of x is logarithmic. This plot better than others shows something

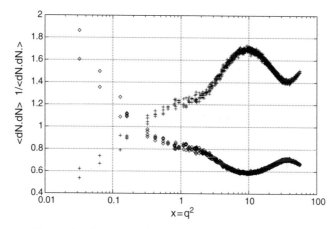

Figure 25. Plot including Large q up and beyond the minimum up to the nearest-neighbor peak. $\bar{C}$ and its inverse $1/\bar{C}$ plotted against $x = q^2$. Semilog scale to show asymptotic region $x \leq 0.4$. The little hump, minimum, and n.n. peak are seen. Small $N = 2880$ system of reverse dimers.

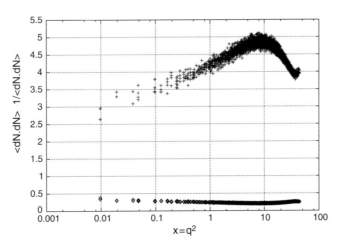

Figure 26. $\bar{C}(q)$ plotted against $x = q^2$ just like in Fig. 25 for chain-like molecules; the minimum and the nearest neighbor peak are there. $\ell = 8, N \simeq 4000, T = 1$.

of a hump of slightly increased scattering; this hump interferes with the steady decrease of $\bar{C}$ with increase in x. This may be the effect of protrusions; these short-wavelength fluctuations ought to be similar to the "hat-like" forms searched for [66]. The hump in this plot is the very largest of all. Fig. 26 shows an identical plot for a bilayer made of chain-like molecules ($\ell = 8$); here the inverse makes the minimum apparent—also near $x = 10$. The nearest-neighbor peak is also present in this plot—but visible near $x = 40$.

VII. DISTRIBUTION OF ORIENTATIONS

The amphiphiles making up the bilayer are anisotropic molecules, and therefore we can inquire about their orientation [39, 61]. In the usual setup, the bilayer is nearly perpendicular to the z-axis of the simulation coordinate system. Its projected area $A_\perp = L^2$ lies in the x–y plane. Therefore the amphiphiles will be, on the average, parallel to the z-axis, but about half will point "upward" (i.e., in the direction of the increasing z) and the other half will point "downward." Therefore we must distinguish the two monolayers making up the bilayer; the "upper" layer is the one for which the average $\langle z \rangle$ is larger than $\langle z \rangle$ in the "lower" layer.

In our model an amphiphile is made up of spheres connected by elastic unbreakable bonds: One sphere is of type "a," standing for the polar "head"; and the remaining ℓ spheres are making up the flexible tail. We choose the first bond

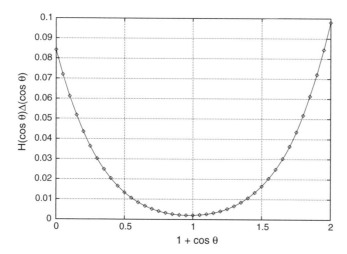

Figure 27. Typical histogram of $\cos\theta$, $H(\cos\theta)\Delta(\cos\theta)$ plotted against $\mu = 1 + \cos\theta \in (0,2)$. The two monolayers correspond to $\mu \in (0,1)$ and $\mu \in (1,2)$. The actual value of H depends on the size of histogram box; all boxes always were equal to $(1/20) = 0.05$.

as the basis for defining the orientation as its angles θ, ϕ made with $+z$ axis for the "upper" monolayer and with $-z$ direction for the "lower" monolayer. Whatever the tail does, the orientation we use is defined only by the first bond, which is an a–b bond. The angle ϕ is averaged over so finally we determine the orientation as the value of $\cos\theta$.

During a simulation run, one easily can collect not only the averages but also the histogram, in this case the histogram of $\cos\theta$, known for each amphiphilic molecule at all times. Such a typical histogram $H(\cos\theta)\Delta(\cos\theta)$ is shown in Fig. 27. The falling part represents the lower monolayer for which $\cos\theta < 0$, and the rising part represents the upper monolayer for which $\cos\theta > 0$. All histograms look quite the same, and only a detailed scrutiny reveals the dependence on the area a and on type of amphiphiles simulated.

The striking regularity of these raw data allows for further processing: calculating averages and following the shifts of a special point of the histogram.

The calculated averages of $\cos\theta$ and of $\cos^2\theta$ are plotted in Figs. 28–31. These averages are calculated for each monolayer separately and are different— sometimes only slightly, sometimes markedly. In order to show the two mono-layers together, we plot one average $\cos\theta$ with a changed sign.

The remarkable feature is the clear change of slope between the extended region EX and the floppy state of the bilayer in FL. These were seen in Section III on

J. STECKI

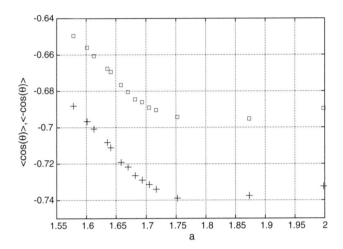

Figure 28. Average $\langle\cos\theta\rangle$ (for each monolayer separately) for a bilayer made of $N = 3980$ amphiphiles with chain length $\ell = 4$ (i.e., 4 beads plus one for the "head") at $T = 1.35$. Boxes: $(-)\langle\cos\theta\rangle$ of the upper monolayer. Note the serious difference between the upper and the lower monolayer. Note the change of slope as imposed a moves the bilayer between the EX and FL regions. There were over 80,000 solvent molecules at this high liquid-like density of $\rho = \mathcal{N} = 0.89$.

lateral tension. The abruptness of this change seems to suggest a transition akin to phase transitions, with these averages as possible order parameters.

For reverse dimers (Fig. 35), all changes of slope appear much smaller and softer.

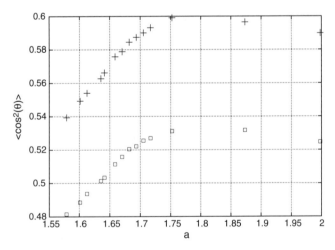

Figure 29. Average of $\cos^2\theta$, for each monolayer separately from the same histogram data as in Fig. 28.

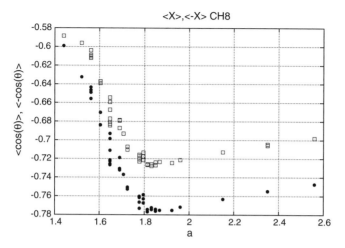

Figure 30. Average $\langle \cos\theta \rangle$ (for each monolayer separately) for a longer chain (tail) in amphiphiles, $\ell = 8$, $N = 450$ (see Fig. 1) at $T = 1$. Comments to Fig. 28 apply. $\mathcal{N} = 14,050$, N is really small; $\rho = \mathcal{N}/V = 0.89$. Filled circles: $\langle \cos\theta \rangle >$ for lower monolayer. Squares: $(-)\langle \cos\theta \rangle >$ for upper monolayer.

Histograms can be fitted readily to a polynomial, taking the lowest point (see Fig. 27) as given; the polynomial of 4th degree gave excellent fits. However, plots of fitted parameters against the area had more scatter and were inferior to the plots shown.

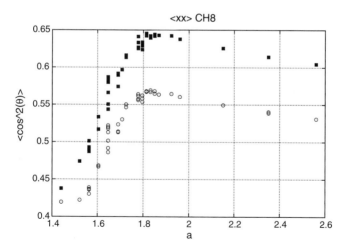

Figure 31. Average of $\cos^2\theta$, for each monolayer separately from the same histogram data as in Fig. 30. See caption to Fig. 30.

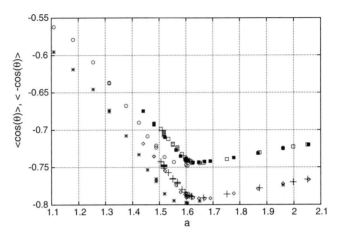

Figure 32. Size dependence of $\langle \cos \theta \rangle$ (for each monolayer separately) for the shorter amphiphile with $\ell = 4$, plotted against the area per head a, for three sizes denoted "s," "m," and "b" with $N = 450$, 1800, and 3980, respectively; all at $T = 1$, $\rho = \mathcal{N}/V = 0.89$. See text. Pluses, "b"; diamonds, "m"; stars, "s". For lower bilayer: squares, "b"; filled squares, "m"; circles, "s"; $(-)\langle \cos\theta \rangle$ for upper bilayer.

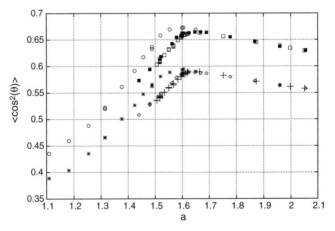

Figure 33. Size dependence of $\langle \cos^2\theta \rangle$ (for each monolayer separately) from the same histograms as in Fig. 32. See caption to Fig. 32.

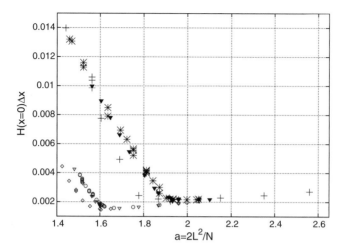

Figure 34. Area dependence of the lowest point of the histogram—that is, of the probability of $\cos \theta = 0 \pm 0.05$. With one data entry per histogram two lengths $\ell = 4, 8$ and three sizes can be shown together. Sizes $N = 450, 1800,$ and 3980 are marked as "s," "m," and "b." Plus signs, 4m; diamonds, 4s; filled circles, 4b; squares, 8s; filled squares, 8m; circles 8b.

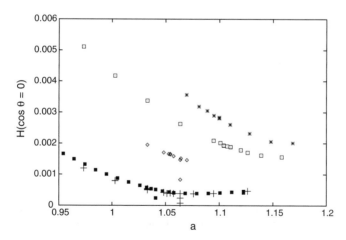

Figure 35. The lowest point of the histogram (see caption to Fig. 34) of $\cos \theta$ for amphiphilic dimers, $\ell = 1$, which are, however, a different model—that is, of reverse dimers. $T = 1.9$. Plus signs, a small system $N = 2880$; filled squares, largest system with $L \sim 100$. In the units of collision diameter of beads, $N \sim 18,000$, Open squares, with long(er) range attractive forces (cutoff $= 3.1$); stars, at a higher temperature $T = 1.9$.

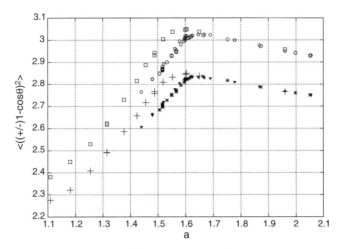

Figure 36. The average of $(\cos\theta - 1)^2$ corresponding to δN^2 is plotted against a for three sizes. See caption to Fig. 34.

A connection with the correlations of surface normals suggests plots of $(\cos\theta - 1)^2$; these are very similar to the plots we have shown (see Fig. 36).

VIII. RADIUS OF GYRATION

The radius of gyration has been invented for polymers, but it is significant for surfaces as well. Whereas for a theory its calculation presents serious difficulties, in any atomistic simulation it can be calculated with greatest ease. The definition is

$$R_G^2 = (1/N_p)\sum_{i<j} r_{ij}^2 \tag{79}$$

where the sum is over all pairs and $N_p = N(N-1)/2$ is the number of pairs. The system made up of N particles (beads, polymer segments) and $|r|$ is the distance between i and j.

In our work, each bilayer isotherm (see Section III) is obtained in a series of simulation runs at different values of a, the area per head; each such run produces one value of R_G. The functional dependence $R_G(a)$ is linear in the EX (extended) regions; a sudden change of slope announces the FL (floppy) region. R_G shares such a behavior—the sudden change of slope—with other quantities (see Section VII) such as the angle averages. In the FL region the linearity of $R_G(a)$ is less perfect at higher temperatures. At low temperatures we find two straight lines crossing at the (presumed) transition point.

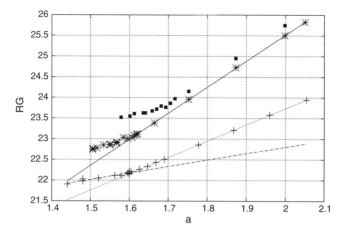

Figure 37. Radius of gyration of a bilayer plotted against the area per head a. Each plot at constant temperature ($T = 1$) except for black boxes ($T = 1.35$). Molecules of chain length $\ell = 8$, ($+$) signs; chain length $\ell = 4$, stars and black boxes. The line joining stars was fitted to the interval (1.6–2.1) in a; all lines are there to guide the eye. In order to show these data together, the set represented by stars was shifted by -0.5 and the set represented by ($+$) signs was shifted by $+6.0$. Otherwise the stars and boxes coincide, quite accurately in the EX region.

Figure 37 shows a few plots of such series $R_G(a)$; the lines are there to guide the eye.

IX. TRUE AREA

The true area of the bilayer has been determined [33, 45, 46] by several methods. But the very concept of the membrane area is tied to the model of a mathematical surface and therefore, starting from the positions of particles in space, some kind of smoothing is necessary. A particularly clear and detailed exposition [45] is available; it appears that the "best" way was to use triangularization. The area L^2 of the x–y plane was filled with an array of hexagonal cells, and the z-coordinate assigned to a cell was the average of the z-coordinates of particles which happened to be inside that cell. As particles, one uses the heads of amphiphiles. The size of the cells is arbitrary and is optimized so as to give consistent results for different areas and also different sizes of the systems simulated. The results thus obtained [40] showed a plot qualitatively similar to plots of $\Gamma(a)$ of Section III, namely a linear relation in the "EX" region of large areas—in this case the $y = x$ relation between the true area and the projected area and a plateau in the FL region. There follows a common (monotonous) curve of Γ versus true area per head, with the exception of the smallest systems with really small $N = 128$ and 288. The common curve is not a straight line, hence the bilayer is a "non-Hookean material."

A different variant [46] was to use a square $\sqrt{N} \times \sqrt{N}$ mesh of cells; the z-coordinate assigned to a mesh point was a weighted average of four neighboring cells. Monotonous plots of Γ against A_{true} were obtained [46].

One can also try to estimate A_{true} from the theoretical predictions of the small gradient theory [66–68] described in Section V. Directly from

$$A = L^2 + \sum_{\mathbf{q}} (1/2) q^2 h_{\mathbf{q}} h_{-\mathbf{q}} \tag{7}$$

calculating the average of $|h_{\mathbf{q}}|^2$ one finds

$$\langle A \rangle = L^2 + (1/2) \sum_{\mathbf{q}} kT / (\kappa q^2 + \gamma) \tag{31}$$

For the case $\gamma = 0$ the sum has been calculated [16], and for better numerical accuracy it is better [45, 46] not to convert the sums to integrals.

Still another method uses the Fourier amplitudes. As these are collected during a simulation run, from which the structure factor will be determined, one can use a finite number of $h_{\mathbf{q}}$'s for inversion to construct $h(x, y)$ and then use (numerically) the exact relation 20. This was apparently used in one instance [33].

There is also a group of methods used in geology, prospecting, and related fields where a representation of Earth surface or some geological patterns is needed—called "spatial data" or "geostatistics." A standard weight is the inverse distance squared (other powers are rare), used when constructing a grid and assigning a "height" to the mesh point. This field is quite apart, unknown to simulators.

X. FLUCTUATIONS OF LATERAL TENSION

Besides the lateral tension Γ itself, it is worthwhile to look at its fluctuation. The standard measure of fluctuations of a quantity X is the average

$$\langle \Delta X^2 \rangle = \langle (X - X_0)^2 \rangle = \langle X^2 \rangle - X_0^2$$

where X_0 is the mean, the average of X:

$$X_0 \equiv \langle X \rangle \tag{80}$$

For the lateral tension Γ we can do better than that, namely we determine the entire probability density $P(\hat{\Gamma}) d\hat{\Gamma}$. This is done by collecting the histogram of Γ. Each histogram is a product of one run—that is, at constant temperature T, density ρ, and given area L^2. For all systems we studied, the histogram invariably followed

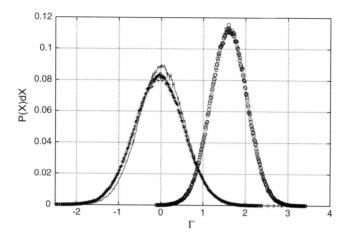

Figure 38. Probability distribution of the lateral tension Γ: in the EX (extended) region (circles), near the transition EX–FL and the tensionless point $\Gamma = 0$ (full line), and deep in the FL (floppy) region [(+)-signs], Large system $N \simeq 4000, T = 1, \ell = 4$.

with greatest accuracy the Gaussian distribution

$$P = P_{\max} \exp[-\alpha(X-X_0)^2] \tag{81}$$

with X_0 equal to the average value of lateral tension for the given run, $X_0 = \Gamma$. The spread of $\hat{\Gamma}$ (i.e., of instantaneous values of Γ) was large—but accurately Gaussian. This was also confirmed by semilogarithmic plots that displayed excellent symmetric parabolas. As can be seen from Fig. 38, at $X-X_0 = -3$ or $+3$ the histogram boxes still were not empty.

All histograms were fitted to Eq. 81; fits of many hundreds of data points to an equation with two constants was very reliable. The quantity $P_{\max}$ is irrelevant because it reflects the size of histogram box, but it disappears on the normalization of the integral to unity. The quantity α is unchanged by the normalization. The width, defined as the distance between the two symmetrical inflection points, is $\sqrt{2/\alpha}$; the value of the Gaussian function at the inflection is $P_{\max} * 2/\sqrt{e}$, where e is the Euler constant. This relation was useful to check the fitted value of α.

Data were obtained along the bilayer isotherms, as explained in Section III, in batches of 10 or more individual runs covering the entire area of the stable existence of the bilayer. We have found two distinct regions, the extended bilayer region "EX" at larger areas and the floppy bilayer region "FL" at smaller areas. We choose to show in Figure 38 three histograms, one deep in the region "FL," one "at" or near the tensionless state $\Gamma = 0$, and one deep in the extended region "EX." The points are raw data, the lines are fitted Gaussians with imposed $X_0 = \Gamma$. The

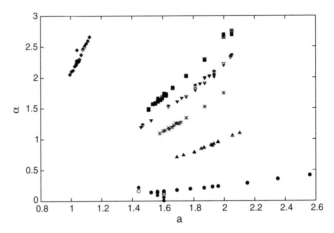

Figure 39. Fitted width-parameter α [Eq. (80)] plotted against the area per head a for a series of bilayers along the respective bilayer isotherms. Circles, up-triangles, and down-triangles: $\ell = 8$, for small, medium, and large $N = 450, 1800, 4000$, respectively. Squares, $\ell = 4, N \simeq 4000, T = 1$; stars, the same at $T = 1.35$; filled diamonds, reverse dimers $N \simeq 18{,}000$, $T = 1.9$, $L \simeq 100$.

distributions are broader for compressed floppy bilayers and sharper for extended (and flatter) bilayers. This is seen in Fig. 39, which collects fitted α for many systems plotted against specific area per head a. Smaller α mean broader Gaussian distribution. The cluster of points at smaller a refers to the reverse dimers; all other clusters refer to bilayers made of chain molecules. A very smooth gradual change of α is to be noted. Incidentally, we have also fitted the same data to an extended distribution which had next two powers of ΔX, but this did not bring anything new: the parameter α was modified quite insignificantly, and the other two constants were of the order of 10^{-3} or less.

The variation $\alpha(a)$ is so smooth that no trace of the existence of two regions can be detected; all points cluster along the respective line, almost a straight line.

There is no visible sign of the transition "FL" $\Leftrightarrow$ "EX" either. Enlarged individual plots of bilayer isotherms do not reveal anything that is not seen in Fig. 39. The only possible very weak sign is the change of slope seen in the smallest system of molecules eight monomers long [marked (CH8)], the same system showing the most dramatic differences between the two parts of the isotherm, as shown in Fig. 1 and elsewhere.

Normally one would expect a raise in fluctuations in the transition region, but nothing of the sort is found. If there is a phase transition, there might be an increase in fluctuation of Γ, just like a peak in heat capacity—that is, an increase in the fluctuation of energy will signal a phase transition.

The behavior of α also is in contrast with other quantities, determined in the very same simulation runs, described in Sections V–VIII.

XI. WHY, IT'S A BUCKLING TRANSITION!

Such was the comment I received from a physicist who was kind enough to give his view. But not much is known about the transition of the bilayer between the extended state "EX" and the floppy state "FL." As we described in Section III, simulations that were not restricted to the tensionless state of either variety choose data points along the bilayer isotherm [33, 37, 39, 42–46]. At fixed temperature T, global density ρ, and fixed number N of amphiphiles, one can still go from one box shape to another, changing the area $A_\perp = L^2$ and the area per head $a = 2L^2/N$. These simulations are in full agreement as to the shape of the lateral tension curve $\Gamma(a)$, schematically simplified to Fig. 6. Only sufficiently large systems comply with that scheme; small and very small systems behave according to Fig. 1. The compressed bilayer originally planar with moderate or small amplitudes of undulations extrudes or "buckles" into the third dimension. The eminent question is the nature of this transition.

Looking for similar situations, we find above all the experiments on vesicles [57] (see also Refs. [58–60]). By applying external forces one finds dramatic (almost discontinuous) change from the extended vesicle to a floppy one—indeed the very expression and the concept of "floppiness" came from the physics of vesicles [55, 65–69]. Reference [57] contains speculation about the nature of the transition.

Another situation is found in the tethered membranes and their simulation [73]. The model is of spheres connected permanently to each their neighbors, fixed once and for all; the bond may be rigid—that is, of fixed length or extensible but nevertheless permanent. There all evidence is of a discontinuous transition, *the crumpling transition.* The term "crumpling" comes about from similarity to crumpled paper. On compression, a piece of paper develops ridges, so does a fender of a damaged automobile, and so look the images of tethered membranes sufficiently compressed. The model of tethered membranes does not apply directy to liquid bilayers, as it involves per se nonzero shear elasticity and the Lamé coefficients are its all-important parameters. MonteCarlo simulations strongly suggested a phase transition [73], but only the heat capacity data supported "the claim that the observed effect is not a mere crossover" [73]. Indeed, very clear peaks bigger with increased system size were obtained [73].

For liquid bilayers the situation is much less clear, really unresolved. The heat capacities C_V are featureless [61]; otherwise there are no reports of the histograms of the lateral tension (see Section X) which could show a visible change on crossing from "EX" to "FL"—in contrast to other quantities such as angles (Section VII, perhaps most convincingly of all), the internal energy $U(a)$ [39], radius of gyration, and fitted parameters of the structure factor (less convincingly) [39].

Lowering the dimensionality from $d = 3$ to $d = 2$ changes the $d-1 = 2$-dimensional membrane into a one-dimensional string now confined to two dimensions (to a plane). Then the model of a membrane becomes identical to a model of polymer; the true area of the membrane becomes the length of the polymer string, and the projected area becomes the projected length onto, for example, the x-coordinate. If we admit a lattice, such as a square lattice, we find the vast world of exact enumerations of which the work of Kumar et al. [74, 75] has many suggestive analogies as it applies "lateral" tension to the last bead of the polymer, having fixed at the origin the first. With a lattice model of a polymer it is not surprising to find discontinuous transitions [74, 75] also of a different kind [76–78]. Whether these analogies will be helpful, the future will tell.

As to the concept of "buckling," it is certainly closer to reality (simulator's reality) than "crumpling," but this term does not properly reflect the foamy and irregular spatial structure of a floppy bilayer which does not—at least on the scale available to computer simulations—display any buckles.

The theoretician view of these "FL" and "EX" states is not uniform. One author [33] dismisses his datapoint on the suspicion (correct as it turns out; see Fig. 1) it might be inside the floppy region or, rather, outside the regular region of interest. Another authority in this field communicated to this author [61] that the bilayer "will transform itself to another form," whereas the theoreticians [73, 79] firmly hold the view that the floppy state is the "natural" state of the membrane whereas the "EX" state is an artificiality created by the experimentalist or simulator. This, it seems, results from believing in the relevance of "persistence length." But the theory has not decided yet whether the membrane folds will make it *softer* or *more rigid* [65, 80, 81].

XII. UNITS USED

All numerical data quoted in this chapter are in Lennard-Jones units. This potential of intermolecular force, u, adds a repulsion term to the London dispersion force term ($1/r^6$); thus we have

$$u(r) = 4\epsilon((\sigma/r)^{12} - (\sigma/r)^6).$$

To be strict, this is the Lennard-Jones 6–12 potential; sometimes other exponents were used. ϵ is the depth of the minimum and $u(r = \sigma) = 0$; hence σ is called the collision diameter. These quantities scale the energy and the distance. Hence the nondimensional "reduced" temperature T^* is kT/ϵ, any length l^* equals l/σ, and intermolecular energy U^* equals U/ϵ. Pressure is $p^* = p\sigma^3/\epsilon$, density ρ^* equals $N\sigma^3/V$, area A^* is A/σ^2, and time t^* equals $t \times \sqrt{\epsilon/(m\sigma^2)}$, where m is the molecular mass. Any surface (interfacial) tension γ such that $\gamma \times area$ is an energy, including lateral tension Γ, is $\gamma^* = \gamma\sigma^2/\epsilon$.

To convert to SI units, usually the mass is taken to be about 36 or 40 atomic units, $m = 40/N_{Av}(g)$ with the Avogadro number $N_{Av} = 6.023 \times 10^{23}$, $\epsilon = (1. - 3.)/N_{Av}(kJ)$, $\sigma = 0.3 - 0.5$ nanometers.

Since no confusion could possibly arise, in the text all asterisks were dropped because all quantities were nondimensional, scaled by molecular units.

There are advantages in using the molecular units of σ and ϵ; for example, the "areaperhead" comes out near 2.0 for chain-like molecules, when the underlying bulk density is near 1 (e.g. 0.89).

On the other hand, an experiment in the real world produces numbers in the SI units; these are not always converted to molecular units because that involves assumptions as to the values of σ and other molecular parameters.

Acknowledgments

The author is greatly indebted to Dr. Soeren Toxvaerd from the University of Copenhagen, for all the knowledge about Molecular Dynamics and simulations. I also thank him for his hospitality there and for long-time collaboration on common projects in the past.

References

1. S. A. Safran, *Statistical Thermodynamics of Surfaces, Interfaces and Membranes*, Addison-Wesley, Reading, MA, 1994.

2. M. Daoud and C. E. Williams, Soft Matter Physics, Springer, Berlin, 1995.

3. J. F. Nagle, *Annu. Rev. Phys. Chem.* **31**, 157 (1980).

4. J. Israelachvili, *Intermolecular and Surface Forces*, Academic Press, London, 1992 and San Diego, CA, 1997.

5. J. S. Rowlinson, *Cohesion*, Cambridge University Press, New York, 2002.

6. R. S. Berry, S. A. Stuart, and J. Ross, *Physical Chemistry*, Oxford Universty Press, New York, 2000.

7. R. Defay, I, Prigogine, A. Bellemans, and D. H. Everett, *Surface Tension and Adsorption*, Longmans, London, 1966.

8. J. S. Rowlinson and B. Widom, *Molecular Theory of Capillarity*, Clarendon, Oxford, 1982.

9. A. Aksimentiev, M. Fijalkowski, and R. Holyst, *Advances in Chemical Physics*, Vol. **121**, John Wiley & Sons, Hoboken, NJ, 2002. See also, e.g., W. T. Gozdz and R. Holyst, *Phys. Rev. E* **54**, 5012–5027, 1996.

10. B. Smit, *Phys. Rev. A* **37**, 3431 (1988).

11. B. Smit, P. A. J. Hilbers, K. Esselink, L. A. M. Rupert, N. M. van Os, and A. G. Szleifer, *Langmuir* **9**, 9 (1993).

12. S. Marrink, M. Berkowitz, and H. Berendsen, *Langmuir* **9**, 3122 (1993).

13. S. E. Feller and R. W. Pastor, *Biophys. J.* **71**, 1350–1355 (1996).

14. S. E. Feller and R. W. Pastor, *J. Chem. Phys.* **111**, 1281 (1999).

15. M. L. Klein, *Biosci. Rep.* **22**, 151 (2002).

16. E. Lindahl and O. Edholm, *Biophys. J.* **79**, 426 (2000).

17. E. Lindahl and O. Edholm, *J. Chem. Phys.* **113**, 3882 (2000).

18. S. J. Marrink and A. E. Mark, *J. Phys. Chem. B* **105**, 6122 (2001).

19. M. P. Allen and D. J. Tildesley, *Computer Simulation of Liquids*, Clarendon Press, Oxford, 1989.
20. D. Frenkel and B. Smit, *Understanding Molecular Simulation*, 2nd ed., Academic Press, San Diego, 2002.
21. S. Toxvaerd, *Mol. Phys.* **72**, 159 (1991).
22. S. Toxvaerd, *Phys. Rev. E* **47**, 343 (1993).
23. M. Kranenburg, Phase Transitions in Lipid Bilayers, Ph.D. Thesis, University of Amsterdam (promotor B. Smit), 2004. Available on the web; this is a minefield of references to computer simulations, especially with DPD scheme.
24. M. Venturoli and B. Smit, *PhysChemComm.* **2**, 10 (1999).
25. M. Kranenburg, M. Venturoli, and B. Smit, *J. Phys. Chem.* **B107**, 11491 (2003).
26. M. Kranenburg, M. Venturoli, and B. Smit, *Phys. Rev. E* **67**, 060901(R) (2003).
27. A. F. Jakobsen, *J. Chem. Phys.* **122**, 124901 (2005).
28. A. F. Jakobsen, O. G. Mouritsen, and G. Besold, *J. Chem. Phys.* **122**, 204901 (2005).
29. A. Imparato, J. C. Shillcock, and R. Lipowsky, *Eur. Phys. J. E* **11**, 21 (2003).
30. G. Illya, R. Lipowsky, and J. C. Shillcock, *J. Chem. Phys.* **122**, 244901 (2005).
31. J. C. Shillcock and R. Lipowsky, *J. Chem. Phys.* **117**, 5048–5061 (2002).
32. A. Imparato, J. C. Shilcock, and R. Lipowsky, *Europhys. Lett.* **69**, 650 (2005).
33. A. Imparato, *J. Chem. Phys.* **124**, 154714 (2006).
34. R. Goetz and R. Lipowsky, *J. Chem. Phys.* **108**, 7397 (1998).
35. G. Gompper, R. Goetz, and R. Lipowsky, *Phys. Rev. Lett.* **82**, 221 (1999).
36. J. Stecki, *Int. J. Thermophysics* **22**, 175 (2001).
37. J. Stecki, *J. Chem. Phys.* **120**, 3508 (2004).
38. J. Stecki, *J. Chem. Phys. Commun.* **122**, 111102 (2005).
39. J. Stecki, *J. Chem. Phys.* **125**, 154902 (2006). (www.arxiv.org) arXiv:0412248 (available as of December 10, 2004).
40. J. Stecki, *J. Phys. Chem. B* **112**(14), 4246 (2008).
41. L. C. Akkermans, S. Toxvaerd, and W. J. Briels, *J. Chem. Phys.* **109**, 2929 (1998).
42. W. den Otter and W. Briels, *J. Chem. Phys.* **118**, 4712 (2003).
43. E. S. Boek, J. T. Padding, W. K. den Otter, and W. J. Briels, *J. Phys. Chem. B* **109**, 19851 (2005).
44. S. A. Shkulipa, W. K. den Otter, and W. J. Briels, *J. Chem. Phys.* **125**, 234906 (2006).
45. W. K. den Otter, *J. Chem. Phys.* **123**, 214906 (2005).
46. H. Noguchi and G. Gompper, *Phys. Rev. E* **73**, 021903 (2006).
47. I. R. Cooke and M. J. Deserno, *J. Chem. Phys.* **123**, 224710 (2005).
48. I. R. Cooke, K. Kremer, and M. Deserno, *Phys. Rev. E* **72**, 011506 (2005).
49. G. Brannigan and F. L. H. Brown, *J. Chem. Phys.* **120**, 1059 (2004).
50. G. Brannigan, L. C. L. Lin, and F. L. H. Brown, *Eur. Biophys. J.* **35**, 104 (2006).
51. O. Farago, *J. Chem. Phys.* **119**, 596 (2003).
52. O. Farago and P. Pincus, *J. Chem. Phys.* **120**, 2934 (2004).
53. F. Schmid, D. Duchs, O. Lenz, and B. West, *Comput. Phys. Commun.* **177**, 168 (2007).
54. O. Lenz and F. Schmid, *J. Mol. Liq.* **117**, 147 (2005).
55. W. Helfrich, *Z. Naturforschung* **28C**, 693 (1973).
56. R. Evans, *Adv. Phys.* **28**, 143 (1979).

57. E. Evans and W. Rawicz, *Phys. Rev. Lett.* **64**, 2094 (1990).

58. M. Bloom, E. Evans, and O. G. Mouritsen, *Q. Rev. Biophys.* **24**, 293 (1991).

59. O. G. Mouritsen and M. Bloom, *Annu. Rev. Biophys. Biomol. Struct.* **22**, 145–171 (1993).

60. J. F. Nagle and S. Tristram Nagle, *Curr. Opin. Struct. Biol.* **10**, 474 (2000).

61. J. Stecki, unpublished.

62. L. D. Landau and E. M. Lifshitz, *Fluid Mechanics*, Pergamon Press, Elmsford, NY, 1959.

63. J. Stecki, *J. Chem. Phys.* **107**, 7967 (1997).

64. J. Stecki, *Phys. Rev. B* **74**, 033409 (2006).

65. W. Helfrich, *Eur. Phys. J. B* **1**, 481 (1998).

66. W. Helfrich and R. M. Servuss, *Nuovo Cimento* **D3**, 137 (1984).

67. W. Helfrich, in Les Houches, Session XLVIII, *Liquids at Interfaces*, Elsevier, Amsterdam, 1989.

68. A. Adjari, J.-B. Fournier, and L. Peliti, *Phys. Rev. Lett.* **86**, 4970 (2001).

69. J.-B. Fournier, Barbetta, *Phys. Rev. Lett.* **100**, 078103 (2008).

70. O. Farago and P. Pincus, *J. Chem. Phys.* **120**, 2934 (2004).

71. J. Stecki, *J. Chem. Phys.* **114**, 7574 (2001).

72. A. Ciach, J. Dudowicz, and J. Stecki, *Physica A* **145A**, 327 (1987).

73. Y. Kantor, in D. Nelson, T. Piran, and S. Weinberg, eds., *Statistical Mechanics of membranes and Surfaces*, World Scientific, Singapore, 1989 and 2004.

74. S. Kumar, I. Jensen, J. L. Jacobsen, and A. J. Guttmann, *Phys. Rev. Lett.* **98**, 128101 (2007).

75. S. Kumar, I. Jensen, J. L. Jacobsen, and A. J. Guttmann, preprint arXiv:0711.3482v1 [cond-mat.], 21 November 2007 (www.arxiv.org).

76. A. L. Owczarek, T. Prellberg, and R. Brak, *J. Stat. Phys.* **72**, 737 (1993).

77. A. L. Owczarek and T. Prellberg, *Phys. Rev. E* **67**, 032801 (2003).

78. D. Marenduzzo, A. Maritan, A. Rosa, and F. Seno, *Phys. Rev. Lett.* **90**, 088301 (2003).

79. K. Wiese, in *Phase Transition and Critical Phenomena*, Vol. **10**, C. Domb and J. L. Lebowitz, eds., Academic Press, New York, 2001.

80. H. A. Pinnow and W. Helfrich, *Eur. J. Phys. E* **3**, 149 (2000).

81. Y. Nishiyama, *Phys. Rev. E* **66**, 061907 (2002).

AUTHOR INDEX

Numbers in parentheses are reference numbers and indicate that the author's work is referred to although his name is not mentioned in the text. Numbers in *italic* show the page on which the complete references are listed.

Adhikari, S., 96(58), *156*
Adjari, A., 187–189(68), 191(68), 193–194(68), 212(68), 215(68), *219*
Akkermans, L. C., 159(41), 162(41), *218*
Aksimentiev, A., 158(9), *217*
Allen, M. P.: 131(81), *156*; 159(19), *218*
Altevogt, P., 72(162), *92*
Amano, M., 96(37), 132(37), *155*
Ananth, N., 95(28), *155*
Andersen, H. C., 50(98–99), *90*
Arasaki, Y., 152(90), *156*
Archer, A. J., 61(129), 67(143), 69(153), 70(143,154), 72(154,159), *91–92*
Ashcroft, N. W., 23(40–41), 26(40), 29(40–41,48), 30(51), 31(63–64), 33(67–68), 55(112), *89–91*
Auguste, T., 96(41), *155*

Baer, M., 95(20,35), 96(20,57–58,62), 97(35), 98(20,38), *154–156*
Baer, R., 95–96(20), 98(20), *154*
Bagchi, B., 62(135), *91*
Baltuška, A., 96(45), *155*
Bandrauk, A. D., 96(40,64), 114(64), 129(80), *155–156*
Barker, J. A., 49(96), *90*
Barrat, J. L., 29(49), 30(49,54), 33(69), *89–90*
Barth, I., 139(86–87), *156*
Baus, M., 4(15), 16(25–27), 22(37–38), 23(25,39), 28(45), 29(25,37), *88–89*
Beier, T., 18(32–33), *89*
Bellemans, A., 158(7), 165(7), *217*
Ben-Nun, M., 95(18,21), *154*

Berendsen, H., 159(12), 173(12), *217*
Berkowitz, M., 159(12), 173(12), *217*
Berry, R. S., 158(6), 160(6), 163(6), *217*
Besold, G., 159(28), 161(28), *218*
Bildstein, B., 39(78), *90*
Bittner, E. R., 95(22), *155*
Bloom, M., 215(58), *219*
Boek, E. S., 159(43), 162(43), 165(43), 215(43), *218*
Born, M., 95(1), *154*
Brabec, T., 96(39), *155*
Brak, R., 216(77), *219*
Brannigan, G., 159(49–50), *218*
Breger, P., 96(41), *155*
Briels, J., 159(42–44), 162(41–43), 165(42–44), 182(42), 187(42), 215(42–43), *218*
Broeckhove, J., 96(48), *155*
Brown, F. L. H., 159(49–50), *218*
Brown, R. A., 30(62), *89*

Cahn, J. W., 3(4), 62(131), 74(170), *88, 91–92*
Capitan, J. A., 48(92), *90*
Carnahan, N. F., 45(85), 50(85), *90*
Carrè, B., 96(41), *155*
Carrera, J. J., 96(43), *155*
Casberg, R. V., 3(12), 33(12), 38(12), *88*
Chakradorty, A., 95(19), *154*
Chan, G. K.-L., 64(141), *91*
Chandler, D., 50(97–99), *90*
Chapman, W. G., 57(121), *91*
Charron, E., 96(63), *156*
Chaudhuri, P., 4(19), *88*
Child, M. S., 95(32), *155*

Advances in Chemical Physics, Volume 144, edited by Stuart A. Rice
Copyright © 2010 John Wiley & Sons, Inc.

SUBJECT INDEX

Advances in Chemical Physics, Volume 144, edited by Stuart A. Rice
Copyright © 2010 John Wiley & Sons, Inc.